CONTENTS

ACKNOWLEDGMENTS

I WOULD LIKE TO THANK the following people. Gil Geis researched the Kemmler case at one time and provided me with material for my initial foray into the subject. Over the years he has been both friend and mentor.

My friend Richard Wright has provided me with his scholarly advice and oversight since we first met in 1978 at the Institute of Criminology, University of Cambridge, and during the almost five years I did commentary for National Public Radio. I'd like to take this occasion to thank him.

My friend James Lawrence is a superb editor who was always available to help me clarify my thoughts.

Dennis Longmire, a professor of criminal justice at Sam Houston University, accompanied me to the execution in Huntsville, Texas.

Marc Kristal is a New York City screenwriter and journalist, whose encouragement, advice, friendship, and own enduring interest in the first electrocution proved invaluable.

Alfred and Timothy Kemmler, the great-grandnephew and great-great-grandnephew of William Kemmler, shared with me their family history and have been extremely supportive of this project.

Bryan Goodwin, the reference librarian at Mount Holyoke College, was able to find many important resources that would have otherwise eluded me.

George Tselos, an archivist at the Edison National Historic Site, graciously gave of his time and knowledge during my many trips to the Edison site during the 1990s. His successor, Leonard DeGraaf, continued the tradition.

Paul Fuller did historical research at the Buffalo Historic Society, providing me with information on Buffalo in the 1880s, and, in a fine piece of detective work, uncovered the transcript of Kemmler's trial, long thought to be nonexistent.

Deborah Denno, a professor at Fordham Law School, read the legal parts of the manuscript. The leading expert on constitutionality of methods of executions, she graciously offered her wisdom and expertise.

Larry Fleischer, attorney and adjunct professor at New York University and City College, offered me his knowledge of late-nineteenth-century criminal law and courtroom procedure in New York State. He was also a constant source of bibliographic information.

My colleague Howard Nicholson, professor of physics at Mount Holyoke College, answered my numerous questions about direct and alternating current.

My editor at Knopf, Jane Garrett, provided the right mixture of personal and professional qualities, prodding, and support that every author needs to see a project to completion.

And to my wife Danielle Federa, on whom not enough appreciation can be bestowed.

I also benefited from several former editors of the *Mount Holyoke News*, who worked for me as student assistants and during the summer. Elizabeth Cook worked as my student assistant and summer researcher when she was a student at Mount Holyoke and later edited the manuscript. Her contribution was especially significant; she offered many helpful suggestions and helped to move the manuscript along. Becky Mazur worked for me as a student assistant and during the summer and did a substantial amount of work for chapter 2 on the lives of Edison and Westinghouse. Sarah Gamble worked for me during the summer as a research assistant and helped with the organization of the book. Kate Schmeckpeper worked for me as a student assistant during the final stages of the manuscript, giving tirelessly of her time and considerable skills. Other Mount Holyoke students were invaluable: Julie Rubin, who helped research and analyze chapter 6 on the evidentiary hearing; Amy Cortright and Katie Herrold, who worked for me as student assistants, were both unusually helpful; Lisa George, who worked for me as a student researcher as well as part of the summer, provided valuable assistance. They are all now former students and current friends.

Pat Kuc made sure that things got to the publisher on time.

Fluffy, an orphan mallard duck, appeared one day alone near the waterfall. We adopted each other and during the summer of 2000 she sat at my feet while I worked on the computer. In the fall she successfully returned to the wild, her job completed.

And, finally, thanks to my twelve-year-old son, Austin, who when just five or six years old would come into my study to offer suggestions and helpful hints, and who, in the last six months, pushed me to finish by remarking out of the blue, "So, Dad, have you finished the book yet?"

INTRODUCTION

MY FIRST ENCOUNTER with the electric chair occurred when I was a small boy growing up in the shadow of the bridge that separates Boston from its northern suburbs. On a hot summer afternoon, my mother took my brother and me down to Bellingham Square to see a full-scale model of the electric chair, and an exact replica of a prison cell. I do not remember whether the state of Massachusetts or the city of Chelsea ran the exhibit, but I will never forget the grim lesson of crime and punishment it was designed to teach: if you broke the law, then this is what would happen to you.

Mother, always fearful that she might lose her sons to the allure of the streets, spent most of the day reinforcing the message of the traveling exhibit: "Crime doesn't pay. They'll get you in the end." She emphasized how most criminals began as small children stealing coins from their mother's purse—a crime I had recently committed—and eventually graduated to big crimes like armed robbery and murder. "Small criminals, small crimes; big criminals, big crimes," she used to say. As we walked up the hill toward our tenement house, my brother and I were filled with fear and apprehension. Would we end up this way? Was this our destiny?

That night both of us had nightmares. I remember being strapped into the chair, the skullcap cutting into my head, the leather mask pressing hard against the bridge of my nose, my ankles throbbing with pain from the pressure of the leather leg straps.

Naturally, I was innocent of the crime that had brought me to this dreadful moment, but somehow the evidence against me was both overwhelming and incontrovertible. I woke in a cold sweat just as the executioner pulled the switch. As I struggled to gain my bearings, my brother burst out of his nightmare. For a long time we just looked at each other. Then, without speaking, we both rolled over and went back to sleep.

In the morning, Mother wanted to talk about the traveling exhibit, desiring to bring home the moral lessons of crime and punishment. My brother and I did the best we could to ignore her, for we had already experienced the ultimate lesson in deterrence. To this day, I cannot entirely shake the feeling of being strapped into the chair, and being condemned to die for a crime the real me did not commit, although my brother seems to have forgotten the incident. I often wonder if I am better off for the experience—if this is what enabled me to escape the lure of the streets and to steer clear of the law. Or has my life been oddly diminished by this haunting image of my own destruction at the hands of the law?

MY SECOND ENCOUNTER with the chair came twenty years later when I was a graduate student at Northeastern University in Boston. My professor, the late Stephen Schafer, asked me to take his undergraduate students on a tour of the maximum-security prison at Walpole. I had never been to prison and I jumped at the chance to see what life on the inside was really like. It had been nearly twenty-five years since Massachusetts executed its last criminal, and condemned prisoners were now housed with the general prison population. Near the end of the tour, a guard asked if we wanted to visit death row. It was a dark, dreary place, made even more dismal by its sudden abandonment. The long corridor of six-by-nine-foot cells that once imprisoned desperate and angry men awaiting execution was no longer filled with anxious life.

Over in the corner was the old electric chair, the one used to execute, among others, Italian anarchists Nicola Sacco and Bartolomeo Vanzetti in the 1920s. To my surprise, the guard asked me if I wanted to sit down. Put on the spot and not wanting to appear cowardly, I obliged. As he strapped me into the chair, I glanced over at the switchboard, checking to make certain that the electrodes were no longer attached. Then, I sat there for a moment and imagined my own execution. On the way out the guard told me that in all the years he had conducted this tour, I was the only one who ever took him up on his offer. For some time afterward, I would quip to friends that I was the last man to sit in the Massachusetts electric chair—and the only one who lived to talk about it.

THE RESEARCH for this book provided my third encounter with the electric chair, although the route it took was anything but direct. For some time I had been interested in the social transformation of punishment—how our ideas about the nature of punishment, especially the death penalty, have changed over the years. With the debate over capital punishment in America at an impasse, with the issues of deterrence, racial discrimination, proportionality, wrongful convictions, and the execution of juveniles at a standstill, I decided that the best way to contribute something new to the public conversation about the death penalty, and to recapture the vitality of the debate that once engaged the best minds in America and abroad, was to focus on methods of execution. To gain a deeper understanding of our culture of punishment, it seemed to me that *style* was suddenly as important as *substance: how* we execute suddenly a better way to understand why and who we execute.

I began my inquiry with the observation that each so-called advancement in the technique of killing has resulted in a temporary increase in the public's acceptance of capital punishment. Historically, new methods of execution develop during periods of mounting opposition to the death penalty. Opponents tend to use the gory details of a botched execution to rally opposition to capital punishment. This was the case with the scaffold, the guillotine, the electric chair, the gas chamber, and, more recently, the hypodermic needle or syringe. Consequently, it has been the proponents of capital punishment—those who wanted to retain the death penalty—who have led the effort to devise a more humane method of taking the life of a condemned criminal. If executions are to continue, the reasoning goes, they must be made palatable to the public.

With this in mind, I began an investigation of lethal injection, the newest and most "humane" method of execution. From my studies on the insanity defense, the medicalization of crime and deviance, and the use of psychological descriptions to explain away political crime, I was aware of how the participation of doctors and the intervention of medical technology into issues of public policy can transform moral, political, and ethical questions into scientific ones. And I realized how difficult it can be for a society to resist easy technological solutions to its most vexing social problems. Indeed, my own attitude toward the death penalty was shaped by exposure to the various methods of execution, as much as by the moral or philosophical arguments advanced for or against capital punishment.

After ruminating over these ideas for many months, I decided that if I were to write about the modern execution ceremony, I needed to witness firsthand an execution by lethal injection. My opportunity came when a colleague from Huntsville, Texas, called to tell me that he had made arrangements for us to witness an execution. Thomas Andy Barefoot, a thirty-nine-year-old former oil field roughneck who was convicted in 1978 for the murder of a Texas policeman, was scheduled to die in two days. Since Barefoot believed that God had forgiven him—that God had already stayed his execution—he had declined to submit a witness list. This left room in the death chamber for the two of us.

The next day, my colleague picked me up at the Houston airport and we drove for two hours in near total silence. As we entered Huntsville at about ten o'clock at night, I was surprised to see that the grounds and stone walls of the prison were brightly lit—like a baseball park. More than two hundred deputies from the Texas Department of Corrections surrounded the perimeter of the prison, a show of force I had not anticipated. There were television trailers in the parking lot and makeshift studios set up on the grass. Two groups of demonstrators were being held back and apart by the police. One group came to protest the cruelty and barbarity of the death penalty. They held candles, prayed, and sang songs of peace and love. The other group, much larger, had come to show their support for capital punishment. They were loud, full of mischief, chanting slogans such as "Fry the Bastard" and "Burn Baby Burn."

Perhaps it was my tweed sport coat and bow tie, but reporters immediately recognized me as an outsider. They crowded around, wanting to interview me. I was so unnerved that all I could say was that I was here only to observe—nothing more, nothing less. Just as we were about to enter the prison, my friend told me that he had decided not to go, that I would have to proceed alone. For a moment I thought about turning back, but I had faced those demons of doubt before boarding the airplane in Boston. And, besides, I had come too far to turn back now.

I entered the prison with a new determination. Although the prison itself was old and sturdy, the reception area had been tacked on more recently, most probably in the 1950s. With its artificial wood-panel walls and dropped ceiling, it looked more like the inside of a cheap motel than a maximum-security prison. A clerk wearing a Department of Corrections uniform escorted me to a small room where a few official witnesses, including the attorney general of Texas, were waiting. The solemn atmosphere reminded me more of a funeral parlor than a prison, although the deceased was not yet dead. One of the men even asked me, "Did you know Mr. Barefoot?"

As we waited, I tried to reassure myself that death by lethal injection was a nonevent. My reading of the history of capital punishment had led me to believe that most of the gore and horror had been taken out of executions. We no longer execute in public, nor do we still regard the infliction of pain and degradation as a necessary part of the ultimate penalty. Thomas Andy Barefoot would drift off to sleep like a hospital patient who had just received general anesthesia. He would die a quick and painless death, if not with dignity, at least without humiliation or horror. Lethal injection had eliminated the gore. But, as I would soon find out, none of the horror.

When we entered the death chamber just before midnight, Barefoot was already strapped to a hospital gurney. Tubes were stuck in both his arms. Saline solution flowed through his veins. He looked up and managed a smile. I had expected that he would be asleep or heavily sedated. He was alert and wide awake. I was frightened.

I had expected to observe the execution, but I did not expect to be observed. The condemned man looked right at me. Only a few feet separated us. I was afraid he would try to touch me. Barefoot did not look like his picture. He was smaller. I had expected a powerful man. The psychiatrists had labeled him a dangerous felon—a prerequisite for execution in Texas—one who was likely to kill again. He looked like a petty criminal. He was dressed in street clothes—no pajamas or hospital sheets, no plastic wrist identification band. He had on shoes.

I was ashamed—ashamed for being there and afraid that he would ask something of me. I was an intruder, the only member of the public who had intruded on this private moment of anguish. In my face he could see the horror of his own death.

The death chamber was brightly lit like a hospital operating room or a television studio. I was so close to him. The brick walls were freshly painted a powder blue. It was air-conditioned cold. The tubes came out of a square hole in the wall. The executioner hid behind a one-way mirror.

Warden Jack Pursley began by asking the condemned if he had any last words. Barefoot strained to hold his head up. Turning toward us, he said, "I hope that one day we can look back on the evil that we are doing right now like we do the witches when they were burned at the stake." After saying that he had prayed all day for the widow of the policeman he had been convicted of killing, Barefoot added, "I want everybody to know I hold nothing against them for anything that they're doing to me. I forgive them all. I'm sorry for anything I've ever done to anybody." Lying there with outstretched arms, he looked oddly like Jesus on the cross.

When Barefoot had finished, Pursley gave the signal: "We are ready." In the adjacent room an unidentified executioner squeezed a syringe containing the lethal drugs, but nothing happened. The intravenous tubes were ten feet long. Barefoot began to talk, asking that we say good-bye to his friends for him. As he began listing the names of several death row inmates, Barefoot let out a terrible gasp. His neck straightened. His eyes bulged and his back arched. He lay stiff on the gurney, glazed eyes fixed on the ceiling, like a soldier standing at attention.

Four minutes passed before the medical examiner pronounced Thomas Andy Barefoot dead. He tried to close Barefoot's eyes, but the lids would not budge. He tried a second time. Still they would not move. Finally the doctor said, "Eyes dilated, respiration stopped, heartbeat slowed. Barefoot is dead." I thought to myself, "No he isn't. His heart is still beating."

All of a sudden we were marching out of the death chamber and down the long corridor of the visiting room. When we reached the outside of the prison the floodlights came on. There were television cameras and microphones. The crowd, mostly students from a nearby university, started cheering. I was not prepared for the celebration. Some people had on Halloween costumes. One was dressed as an executioner, with a black silk hood pulled over his head. He was holding a cardboard ax.

The reporters peppered us with questions. They wanted to know every detail. How long did it take for Barefoot to die? Did he suffer? Did he admit killing the policeman? Or did he maintain his innocence to the end? Was the execution humane? They even wanted to know if justice had been served that October night in Huntsville.[1]

A FEW WEEKS after I returned from Texas, a colleague asked me to lecture on the application of new medical technology to the science of killing. He knew I had witnessed an execution by lethal injection and thought I might be able to bring some special insight to the problem of capital punishment. To prepare myself, I researched the history of lethal injection. Much to my surprise, I discovered that the idea of putting condemned criminals to death with a lethal overdose of drugs had been around at least since the time of Socrates, and the hypodermic needle was first used extensively during the Civil War. Moreover, in 1886–88, New York State established a commission to recommend the best method of executing criminals—and lethal injection by an overdose of morphine or prussic acid came in second behind electrocution. It turned out that lethal injection was not a high-tech death after all, but a low-tech death that had been proposed and rejected more than a century ago. More important, I stumbled upon the story of the bitter struggle between Thomas Alva Edison and George Westinghouse Jr. for control of the emerging electrical-power industry that set the stage for the invention of the electric chair.

· · ·

DURING THE QUARTER CENTURY between the end of Reconstruction and the assassination of President William McKinley in 1901, American society experienced unparalleled social, economic, and scientific progress. The Gilded Age was a time of unwavering faith in science and technology. Things were suddenly possible that never were before: rapid travel by railroad, long-distance communication via the telegraph and telephone. Together they had a tremendous effect on people's lives, perhaps even greater than the social transformation produced by the current information age. Science and technology promised to make life better and death less painful.

It was also a period of stark contradictions. Religion was of utmost importance, but Charles Darwin's theory of evolution was beginning to take hold. Major corporations began to dominate American business. The railroads, steel, oil, and electrical industries captured the public's imagination as never before; people welcomed this development but also feared its unknown consequences. While science and industry promised to transform lives for the better, unfettered corporate growth and rapid urbanization caused unprecedented social and economic upheaval. The gap between the rich and the poor widened as the nascent corporate culture embraced the ideology of survival of the fittest, along with its callous disregard for the poor and less fortunate.

The Gilded Age was not just an age of invention, but of inventors. Inventors were more than heroes, they were national symbols. To the common man, electricity was a powerful and mysterious fluid, not readily understood. In the dexterous hands of the expert, however, it had extraordinary power. No one personified the mystique of the inventor more than Thomas Edison. He was a midwesterner who seemed to embody the best elements of the past and the great promise of the future; he was at once old-fashioned and modern. Edison epitomized the inherent capability of the average person who, through practicing the virtues of hard work, discipline, ingenuity, and especially perseverance, could harness the power of the future.

An instinctive entrepreneur, Edison focused on useful inventions that could be marketed to the public. Invention was his business. He had little patience for theoretical speculation that lacked immediate practical application. His emphasis was always on inventions that could be manufactured for wide distribution. Like his friend Henry Ford, who grasped the economic advantages of building a car his assembly-line workers could afford, he sought to profit from improving the life of the common man. The 1877 invention of the phonograph marked the real emergence of the Edison legend. His new invention allowed people to preserve the past, even while he moved them toward the future. Two years later, his light-bulb turned night into day. These inventions became part of everyday life, and Edison himself became a national treasure.

As America moved toward an uneasy embrace of scientific advancement and mass production that promised to make life better, the populace grew nostalgic for a simpler past. Edison seemed to link the familiar world of essential virtues to the new one of infinite possibilities. His lack of formal education only increased his appeal. A self-made man, he toiled in a small, rural New Jersey town to build a new model of science and industry for an increasingly urban landscape. Edison was also a master of public relations, expertly crafting his public image as the quintessential American hero. Even today, Edison remains a national icon, his lightbulb a symbol of American genius. And although it is not generally associated with him, his electric chair remains the symbol of capital punishment.

WHAT FOLLOWS is the story of how Thomas Edison helped influence the state of New York to replace hanging with the new scientific and "humane" method of electrical execution. It reveals how a bold, almost fearless faith in science, technology, and especially progress joined powerful economic interests and personal ambition to produce a change in the method of capital punishment. By unmasking the "story behind the story," I aim to demonstrate how our most cherished social values can be manipulated to serve pecuniary interests: the way in which public policy is affected by behind-the-scenes maneuvering of powerful and often ruthless business interests. The darker side of human nature, the desire for fame and fortune, and a fierce willingness to do whatever it takes were the main factors that motivated Edison's involvement with the electric chair. Far from representing an enlightened humanitarian concern for the welfare of the condemned, the electric chair was invented so that one major electric company (Edison) could maintain its competitive advantage over another (Westinghouse).

IN 1890, a Buffalo huckster named William Kemmler became the first person put to death by electricity. Although the execution itself was terribly botched, supporters proclaimed it a huge success, a triumph of science and humanity over barbarism and brutality. A detailed examination, however, reveals that the invention of the electric chair was not motivated primarily by a humane desire to provide the condemned with a quick and painless death, but by the desperate need of one major electric company to discredit another.

In 1882 Thomas Edison launched the "age of electricity" by illuminating the commercial and financial districts of New York City near his Pearl Street power station. Four years later, George Westinghouse, the inventor of the air brake, illuminated Buffalo, New York. Edison lighted New York City with direct current, or DC. Westinghouse employed alternating current, or AC. This simple distinction holds the key to understanding the origin of the electric chair.

By 1888, as the superiority of alternating current became generally recognized, the Edison Electric Light Company began to lose market share, and a bitter struggle to control the emerging electric power industry ensued. Determined to maintain dominance in the industry he created, Edison tried to frighten the public by claiming that AC was too dangerous for commercial or residential use. Edison used his prestige as an inventor without equal to help persuade the state of New York to electrocute its condemned criminals using an alternating current supplied by a Westinghouse generator.

George Westinghouse was not pleased. To prevent alternating current from becoming known as the "executioner's current," Westinghouse hired a team of New York lawyers to prevent Kemmler's execution. Led by former congressman W. Bourke Cockran, the legal team filed a writ of habeas corpus, challenging Kemmler's electrocution as cruel and unusual punishment under the provisions of the New York State Constitution, as well as under the Eighth Amendment of the U.S. Constitution. At an evidentiary hearing, Edison testified that death by electricity was quick and painless, provided it was accomplished with alternating current. Both the Supreme Court of New York and the New York Court of Appeals turned down Kemmler's petition. Kemmler appealed to the U.S. Supreme Court. Although it refused to decide Kemmler's claim that electrocution was cruel and unusual punishment, courts have continued to rely on *In re Kemmler* for the proposition that electrocution is permissible under the U.S. Constitution.

On August 6, 1890, William Kemmler became the first person to die in the electric chair. At the crack of dawn he received two applications of 1,300 volts of alternating current. The first lasted for only seventeen seconds because a new, unstretched leather belt was about to fall off one of the secondhand Westinghouse generators. Initially, Kemmler appeared dead, but suddenly he began to breathe again. A second jolt was applied that lasted until the smell of burning flesh filled the room, about four minutes. After his charred body stopped smoldering, Kemmler was pronounced dead. Edison must have bribed a reporter because a headline in the morning paper read: "Kemmler Westinghoused."

THE ORIGIN of the electric chair has broad implications for our understanding of the relationship of science and technology to social policy, and especially to the ongoing debate on capital punishment. Kemmler's electrocution is generally regarded as the first modern execution. It was the first to rationalize the process of putting a person to death by employing science and technology in service to humanity, and it was the first time the press was legally barred from attending an execution and publishing its gruesome details. Once stripped of its symbolic content and instructive purpose, the goal of an execution was merely to kill the condemned as quickly, painlessly, and efficiently as possible. Thus, the electric chair embodied two of the most powerful cultural components of late-nineteenth-century America: a fearless, almost fanatical, faith in progress and a growing abhorrence of inflicting physical pain on fellow human beings. While our faith in science and technology has tempered somewhat, our distaste for inflicting pain or mutilating the body remains the driving force behind our never-ending search for a more humane method of execution.

A NOTE ON TERMINOLOGY

WILLIAM KEMMLER was the first person ever to be electrocuted. The actual words "electrocuted" and "electrocution," however, were rarely used by his contemporaries. These terms came into use only after Kemmler's execution, and after much experimentation.

Once New York State had decided that it would execute its condemned prisoners by electricity, letters poured in to popular publications with suggestions about exactly what term should be used to designate an execution by electricity. The *New York Times* called the search for an appropriate word "a desideratum," and newspapers published letters submitting the terms "Browned," "Gerrycide," and "Westinghoused" after prominent players in the origin of the electric chair. Edison's lawyer wrote to his legal partner with an elaborate explanation of what he considered to be the correct etymology:

> I think that the words which Mr. Edison suggests, namely, "ampermort," "dynamort" and "electromort" are all open to the same objection, as proposed words to signify the act of execution by electricity. The trouble lies in the termination. The termination "mort" does in fact come from the Latin *mors*, but carries with it the signification of death only, that is to say, simple, passive death, with no idea involved in it of being put to death . . . I have formed a word from the word *electrica* (which is the Latin form of the Greek meaning "amber") and the word *cædo*, *cædere* meaning "to kill," and I propose the word "electricide" as a word most nearly signifying "the act of killing by electricity." . . . The only other word I can think, under the circumstances . . . is "westinghouse" . . . There is a precedent for it, too, one that could not be more apt or authoritative. We speak of a criminal in France as being guillotined, or condemned to the guillotine. Each time that word is used it tends to perpetuate the memory of and services of Dr. Guillotine, who afterwards died by the same machine he had invented. The adoption of the word westinghouse for a like purpose will go far towards rebutting the claim that Republics are less grateful than Empires.

Included in the numerous suggestions for death by electricity were: *ampermort, browned, electrocide, electrolethe, electrical execution, electromort, electrohanabia, electro-cremation, electrothanasia, electricide, electrodited, dynamort, gerrycide, judicial lightning,* and *westinghouse.* Ultimately, however, the now familiar phrase "electrocution" prevailed, despite the *New York Times* lament, "We pray to be saved from such a monstrosity as 'electrocution,' which pretentious ignoramuses seem to be trying to push into use . . . The Bowery slang makers are likely to hit upon a better word than that, if time is given them." But that did not happen. While the outside world hunted for exactly the right term, death row inmates were content to refer to the chair as "Old Sparky," and the process itself as being "fried."

Executioner's
Current

(National Police Gazette, 23 August 1890)

THE FIRST ELECTROCUTION
William Kemmler was electrocuted in Auburn Prison
for the murder of Tillie Ziegler on August 6, 1890.

CHAPTER ONE

"William, It Is Time"

I N THE PREDAWN HOURS of August 6, 1890, twenty-seven men of law, science, and medicine left their lodgings at the Osborne House and quietly made their way down State Street toward Auburn Penitentiary. It was a dull and gloomy morning with a few wet clouds in the sky. The night before had not been an easy one, and the results were written on the faces of each hunched figure. As the men walked, there was little conversation. Nearing the prison, they encountered a crowd of nearly five hundred spectators. Every tree and rooftop surrounding the ivy-covered stone prison was filled with expectant faces, and young men and boys were perched atop telegraph poles, eager to catch a glimpse of the condemned man scarcely visible through the narrow window of his lighted cell. Western Union had opened a temporary office across from the penitentiary in the dimly lit freight room of the New York Central railroad station.[1] Inside, newspapermen and telegraph operators anxiously waited to dispatch word around the world that the first execution by electricity had taken place.

Although a ticket of admission had been issued to each witness, the men had difficulty gaining entrance to the prison. The crowd was reluctant to give way and security was tight. Warden Charles Durston had ordered the gatekeeper not to let anyone in without a ticket; one witness who had forgotten his was forced to return to the hotel to fetch it. Even the morning shift of guards was not permitted to enter the prison until the bells rang, signifying a completed execution. District Attorney George Quinby, who had prosecuted the condemned man, looked pale as he walked through the prison gate below the statue of the Continental soldier standing guard on its roof. Though he had convicted many murderers, he had never witnessed an execution.[2]

Once inside, the men were escorted into Warden Durston's office where prisoners in white caps and aprons served them coffee and sandwiches.[3] Warden Durston did not join his distinguished guests for breakfast.[4] He had gone directly from his prison lodgings to the basement cell of William Kemmler, the condemned man. After an exchange of pleasantries, Warden Durston drew an official, impressive-looking document from his breast pocket, for the law required that the death warrant be read prior to execution. For his part, Kemmler remained outwardly calm. "William, it is time," said the warden. "I am ready, Mr. Durston," the condemned man responded.[5] Then the warden, his voice trembling, read the death warrant. It differed from all previous warrants only in the prescribed method of execution. William Kemmler was to receive a current of electricity sufficient to cause death.[6]

Kemmler listened in resolute silence. When Warden Durston finished, the condemned man replied: "All right, I am ready."[7] The two men then sat on Kemmler's iron bunk and spoke for a few moments before Durston returned to his office on the second floor. In the entranceway he met the witnesses, now standing about, waiting. They exchanged polite nods, nothing more. The atmosphere was decidedly funereal, although the condemned man was not yet dead.

DURING THE PREVIOUS AFTERNOON, Warden Durston had shown the witnesses the newly constructed death chamber, where electricians were putting the finishing touches on the execution apparatus. Dr. George E. Fell, a Buffalo professor who had played an important role in the chair's final design, volunteered to be strapped into it for purpose of demonstration.[8] As he did so, Warden Durston declared his utmost confidence in the reliability of the chair.[9] Not everyone shared this optimistic assessment: specifically, some experts had doubts about the strength and dependability of the first-ever electric chair, but ultimately, they decided it was too late to make changes. The chair, they claimed, was indeed the perfect example of science employed for the betterment of humanity— death would be quick and painless.[10]

FROM THE TIME of his arrest for the murder of his paramour, Matilda "Tillie" Ziegler, on March 29, 1889, until four days after he was sentenced to death, Kemmler remained in the Erie County jail at Buffalo. Then, in accordance with the law, he was transferred to Auburn State Prison.[11] During the trip, Kemmler told his keepers that years before an elderly fortune-teller in Philadelphia had foretold his execution, and everything had transpired exactly as she had predicted. Five days later, on the night of May 23, Kemmler was placed in solitary confinement. He was allowed no visitors except his keepers, his lawyers, his religious advisers, and a few friends of the warden.

About three months prior to his execution, Kemmler dictated his last will and testament to the head turnkey, guard James Warner. The men were interrupted several times by the sound of hammering from two convicts working nearby on the plain pine box that would serve as Kemmler's coffin. Electrician Edwin F. Davis, who would continue as an executioner for the next twenty-four years—throwing the switch on 240 condemned men—could be heard installing the execution apparatus in the room adjacent to Kemmler's cell. Seemingly indifferent to his impending execution, Kemmler assigned his meager belongings with great care. He designated that his principal keeper, Daniel McNaughton, should receive a pictorial Bible that had provided great solace to Kemmler. To the Reverend Dr. W. E. Houghton he left a pig-in-the-clover puzzle. He gave a slate with his autograph to prison chaplain Horatio Yates. To keeper William Wemple he left a small Bible. To Mrs. Durston, the warden's wife, Kemmler gave fifty autographed cards. While in confinement, Kemmler had learned to write his name. He was very proud of this accomplishment, and presumably wanted to share it with the woman who had taught him. He also believed that after his death, the cards would be of significant monetary value.[12]

Kemmler distrusted reporters and had always refused to answer their questions. But he was talkative with visitors, and at times quite entertaining. "He sings, cracks jokes, and . . . tells stories—the sort of stories that wouldn't look well in print," reported the *Buffalo Evening News*. The paper stated blithely that there was a second reason the ax murderer should be executed: "He is a bad rhymester."[13]

While in jail Kemmler composed several drinking songs, such as this one, published in the *Buffalo Evening News*.

> I used to live in Buffalo,
> The people knew me well,
> I used to go a-peddling,
> A plenty did I sell.

(National Police Gazette, 23 August 1890)

Preparing Kemmler's head for the skullcap

My old clothes were ragged and torn,
  My shoes wouldn't cover my toes.
My old hat went flippity flap
  With a schuper to my nose.

I can't sing sing,
  I won't sing sing,
I'll tell you the reason why.
  I can't sing sing,
  I won't sing sing,
For my whistle is getting dry.[14]

Despite the condemned man's penchant for ribaldry, Sheriff Oliver A. Jenkins reported that Kemmler was a model prisoner. Head turnkey Warner credited Kemmler with "bearing up wonderfully well" under the constant strain of imminent death.[15]

WHEN HE ARRIVED at Auburn on May 24, 1889, Kemmler was an habitual cigar smoker in poor physical condition, and of a "morose" and "taciturn" disposition. While at Auburn, however, Kemmler's personal health and appearance improved greatly.[16] He attributed this change to his enforced temperance and claimed that while in the Erie County jail awaiting trial, the guards had constantly given him tobacco and whiskey. In the months prior to his execution, he was neatly dressed and his collar was turned up in accordance with the latest fashion. His whiskers were trimmed and parted at the chin in the English style, and he always wore a black tie.[17] During his long confinement, Kemmler's health remained good, except for a brief bout with dysentery.[18] He passed his time by singing banjo songs with fellow death row inmate Frank Fish, and listening to his keeper read Victor Hugo's *Les Misérables.*[19]

During Kemmler's long wait on death row, the warden's wife took a special interest in him. Mrs. Durston spent considerable time with Kemmler, singing, reading Scripture, or praying with him. By the end, Kemmler appeared a better man thanks to her intervention.[20] Under her guidance he had learned to read, and seemed to have made peace with his Maker. In a way, all the loving attention devoted to him was compensation for being the first victim of a deadly scientific experiment.[21] The last time she spoke with Kemmler, Mrs. Durston told him he was going to a better place. She took his hand and said, "God be with you, be brave, be strong; everything will come out right."[22]

Mrs. Durston left town for New York City the next morning. Upon her arrival, a friend met her at the railroad station and took her to a country home in Lawrence, on Long Island. Her long association with Kemmler and her genuine concern for his well-being made it too difficult for her to remain in Auburn during the execution. The citizens of Auburn, kept in the dark as to the exact date of Kemmler's execution, naturally took Mrs. Durston's departure as a sign that the fatal day was at hand.[23]

There remained only one possible source of delay. Some observers speculated that the electrical manufacturing giant George Westinghouse might, at the last minute, get an injunction to prevent the use of Westinghouse dynamos in the execution. Attorneys for the Westinghouse Company had fought Kemmler's execution on constitutional grounds right up to the New York Supreme Court, but since the U.S. Supreme Court refused to hear its appeal, there seemed little chance that an injunction would be granted.[24] Although the Westinghouse attorneys had argued that electrocution was cruel and unusual punishment, their real concern was the inevitable association the public would make between death by electricity and its energy source, the alternating current produced by Westinghouse dynamos (generators).[25]

To this end, Westinghouse's attorneys filed a writ for the return of the company dynamos. The writ of replevin claimed that Harold Brown, the electrician who prepared the execution apparatus, fraudulently obtained the dynamos. Westinghouse attorneys argued unsuccessfully that Westinghouse retained an interest in its dynamos even after they were sold, giving them control over any potential misuse.[26]

Initially, Kemmler was sentenced to die during the week of June 24, 1889.[27] To allow his attorneys time to pursue appeals, Kemmler was twice granted a stay of execution. Finally, on July 11, 1890, his legal means of recourse exhausted, William Kemmler arrived in a Buffalo courtroom to hear the sentence of death pronounced for the third time. Accompanied by his keeper, Daniel McNaughton, Kemmler was dressed in an impeccably pressed gray suit and a black derby hat. While confined at Auburn, his dark brown beard and mustache had grown thicker. During his trial, he looked listless and confused. On this day, he was alert, yet he appeared unconcerned. Judge Henry Childs of the New York Supreme Court asked the assistant district attorney if he had any business before the court. "If the court please," he said, "we desire to move sentence on William Kemmler."[28]

Kemmler was asked to stand up, and Judge Childs informed him that all attempts to save his life had failed. "Yes, sir," Kemmler replied. "Have you anything to say, why a time should not be fixed for carrying out the sentence previously passed upon you?" the judge asked. After a long pause, Kemmler responded, "No, sir." The judge said he hoped the prolonged delay had given Kemmler time to think about the enormity of his crime and the justice of his conviction. He then said that the sentence would be carried out during the week of August 4, 1890. The judge concluded with the time-honored phrase, "May God have mercy upon your soul."[29]

Upon his return to Auburn prison, Kemmler appeared to be in good spirits, even though he had less than two weeks to live. He no longer spent as much time with his Bible, but preferred to have novels read to him by keeper McNaughton. Kemmler often made the cryptic remark that he "would never die by electricity"; apparently he still hoped for a last-minute reprieve.[30] Along with fellow death row inmate and convicted murderer Frank Fish, who occupied the cell adjacent to Kemmler's, he joined in the daily activities with the "utmost relish." Like Kemmler, Fish was also confident that he would not die in the electric chair. Fish seemed to be in extraordinarily good spirits, even though, at best, a commutation from the governor would still mean that he would spend the rest of his life in prison.[31]

Until this point, it looked as if Joseph Chapleau, a poor French-Canadian farmer who had beaten his neighbor to death with a billet of wood, would be the first to go. However, on July 16, 1890, just four days before his scheduled execution, Governor David B. Hill commuted Chapleau's death sentence to life imprisonment. The electric chair had been installed at Dannemora Prison, the place of Chapleau's confinement and scheduled execution. Now it would have to be dismantled and moved to Auburn. William Francis Kemmler would go first.[32]

THE NEWS that Kemmler would be first came as a shock to prison officials at Auburn. It was generally known that Warden Durston did not want to preside over the first electrical execution and would avoid it if possible. Durston had hoped to benefit from the experience gained at Dannemora, and he did not want to conduct what many regarded as a risky scientific experiment.[33] Indeed, when Durston traveled to New York City four days before the execution, most people believed he went to persuade Governor David B. Hill that Kemmler had gone insane, requiring postponement of the execution.[34] When Durston returned to Auburn, he remained in his office, refusing to talk with anyone.[35]

The impending execution of William Kemmler was the major topic of conversation among the citizens of Auburn. The town was full of rumors concerning Kemmler's fate, the most significant being that Kemmler had lost his mind.[36] If this were true, it would entitle him to a stay of execution and a commutation of his sentence to life imprisonment. However, the general impression was that Kemmler's lawyers had invented the ruse to save him from the executioner's current.[37] According to his keepers, Kemmler's physical and mental condition was about what it had been on the day he was arrested for the murder of Tillie Ziegler. Jailer McNaughton, annoyed by the persistent rumors, remarked that despite his year on death row, Kemmler was "not a gibbering idiot, nor a driveling imbecile."[38]

Four months before his death, under Mrs. Durston's guidance, Kemmler had become religious and was baptized into the Methodist faith. The night before his execution the condemned man received the sacrament of the last supper. His spiritual advisers felt they had done everything possible for a man of Kemmler's "meager intelligence." Warden Durston joined the ministers and the prisoner. Kemmler said he felt fine and would not "flinch at the end. I am not afraid, Warden, so long as you are in charge of the job. I won't break down if you don't."[39]

On the eve of Kemmler's execution, fellow death row inmate Fish played the banjo and the two prisoners sang "My Old Kentucky Home" and "Wait 'Til the Clouds Roll By" one last time.[40] Listening to the minstrelsy, his keepers were awed by Kemmler's spirit. He showed no sign that death was waiting just outside his door.[41] At 9:30 p.m., the lights were dimmed, and Kemmler was told it was time to end the music. Before returning to his cell, Fish took Kemmler's hand and said, "Keep your courage up, Kemmler, it will all be over soon. I will follow you after a little while." Kemmler muttered, in response, "I guess I will behave all right. It can't come too soon for me. Being so near the end is as bad as the actual going."[42] The two men then looked each other in the eye, paused for a moment, and parted company.

Kemmler spent the rest of the night alone in his cell, aware that at 6:00 a.m. he would be taken to the death chamber. The horror of his predicament seemed not to register with him; outwardly, he remained calm. He slept soundly. Perhaps it was a matter of pride, or maybe a blind faith in the men of science who had promised him that death by electricity would be quick and painless, but Kemmler assured his keepers that he would "die like a man."[43]

IN ORDER to conceal the identity of the executioner and to spare the condemned man and the witnesses from having to watch the switch thrown, Warden Durston had moved the death chair from its initial location to a room immediately under the main stairway.[44] The original execution room had been pronounced perfectly suitable only four months ago, and the change in location came to be viewed as evidence of the bungling and delay that characterized Warden Durston's handling of events. The new room, known as the keeper's mess room, was only a few feet from the room first chosen. Nonetheless, some feared that this last-minute removal of the chair and the reconnecting of its elaborate switchboard might introduce possible glitches.[45]

Edwin F. Davis, a New York electrician who had wired the chair, arrived just before dawn with a replacement voltmeter. Then the work of reinstalling the chair began. In his nearby cage, Kemmler could hear the workmen's conversation and banging. The condemned man listened intently for several minutes, and then said to McNaughton, "They're getting ready, Daniel." When McNaughton told him that the workmen were merely moving the chair from one room to another, Kemmler seemed reassured.

The voltmeter, lamps, and switchboard remained in the old execution room. In the new death chamber, the chair was placed at the lower end near a door leading into the control room. It was in this room that Warden Durston would signal for the current to be turned on. Electrician Davis had arranged an elaborate code of signals with the dynamo room, where C. R. Barnes of Rochester was in charge. An electric bell was to be used; two strokes meant to start the dynamo, and each succeeding double-stroke meant to increase the machine's velocity. A single stroke was the signal to stop.[46]

The chair, constructed of heavy oak with a high, sloping back and broad arms, was bolted to the floor. Heavy leather straps with sturdy buckles were attached to the sides and arms of the chair. The electrode, which passed through a wooden figure four, fastened to the back of the chair and adjustable to any height, was suspended from a horizontal arm attached to the back of the chair, above the headrest. A rubber cup containing a natural sea sponge hung above the head of the chair, and another cup and sponge were attached to its lower back. From these cups two wires ran out

(National Police Gazette, 23 August 1890)

Strapping Kemmler in the chair.

the window and across the roof to the dynamo in the northwest corner of the prison, about one thousand feet away.[47]

The Westinghouse dynamo was driven by a 45 horsepower engine.[48] Two large wires connected it to the chair through a window in the control room. The switchboard, about five feet long and four feet wide, contained a voltmeter, a resistance box, a board with twenty-one incandescent lamps, an ammeter used to measure the quantity of electricity, a regulating switch, and the execution lever itself.[49] The purpose of the lamps was to measure the force of the current, should the voltmeters malfunction. By arranging the lamps in series, the total voltage across the mains could be estimated. Since the lamps required a known voltage, a drop in voltage would be visible if the lamps dimmed.[50] To test the entire execution apparatus and to assure that everything was in working order, a "gaunt, worn-out horse" had been electrocuted the day before.[51]

WARDEN DURSTON returned to the condemned man's cell with Erie County undersheriff Joseph C. Veiling, Kemmler's Buffalo keeper. Kemmler seemed pleased to see Veiling, and insisted that the jailer stay for breakfast. While the two men waited for Kemmler's last meal to be prepared, Reverend Houghton of the First Methodist Church and prison chaplain Reverend Yates entered the cell. Kemmler talked with them at length about his imminent execution, and afterward, Kemmler and the ministers knelt in prayer.[52]

Breakfast having been completed, Veiling told Kemmler that it was time to shave his head, for the skullcap required a direct contact with the scalp. While Veiling cut, the two men talked. "They say I am afraid to die, but they will find that I ain't," said Kemmler. "I want you to stay right by me, Joe, and see me through this thing and I promise you that I won't make any trouble." Veiling tried to reassure Kemmler, but the jailer was himself filled with fear and apprehension. By the time the two men finished talking the straight razor had done its work, and the top of Kemmler's head was bleeding slightly.[53]

At last the time had come. Warden Durston reentered Kemmler's cell. After bidding farewell to Veiling and McNaughton, who had been his faithful companion and spiritual adviser during his sixteen-month confinement, Kemmler glanced around the grim, rusting steel walls of his cell for the final time. "Come, William," the warden prompted.[54]

Walking slowly down the long, narrow passageway, they reached the doorway to the death chamber at 6:38 a.m. Warden Durston entered first, followed by the condemned, and then the two clergymen. The electric chair stood at one end of the small, gray, square room. William Kemmler was an average-sized, broad-shouldered thirty-year-old man with a full beard; before it was sheared, his carefully arranged dark hair had been stylishly clustered around his forehead. He would be executed dressed in store-bought clothes—a sack coat, a dark gray vest, yellow trousers, and white shirt with a bow tie.[55] Kemmler also wore a breastpin Mrs. Durston had given him; it was in the shape of a star, and he treasured it because it was "pointing to heaven."[56]

Kemmler walked into the death chamber in a composed, easy manner. A dreadful silence fell over the assembled men of law, science, and medicine, many of whom looked more anxious than Kemmler.[57] The silence was broken by Warden Durston. "This is William Kemmler. I have warned him that he has got to die and if he has anything to say he will say it."[58] Kemmler bowed before speaking, "Gentlemen, I wish you all good luck. I believe I am going to a good place, and I am ready to go. I want only to say that a great deal has been said about me that is untrue. I am bad enough. It is cruel to make me out worse."[59] After bowing a second time, Kemmler turned his back to the witnesses, took off his coat, and handed it to the warden. Then he straightened his black-and-white bow tie.[60]

Warden Durston stepped forward, and Kemmler, in compliance, sat down on the chair. Sun from a small raised window shone brightly on his face. The warden motioned Kemmler to stand. He needed to check Kemmler's clothing to make certain that it had been cut away to allow for a clean contact between the electrode and the spine. The warden discovered that the sack coat had been cut, but not the white shirt. Taking a pen knife from his pocket, Durston cut two small triangular pieces out of the shirt.[61]

Kemmler sat down again. The warden placed a skullcap on his head. The witnesses muttered as Kemmler calmly turned toward Durston and said, "Now take your time and do it right, Warden. There is no rush. I don't want to take any chances on this thing, you know."[62] The warden then strapped Kemmler into the heavy oak chair. The straps were necessary not only to discourage a possible escape attempt, but to anchor his body to the chair after the electricity was discharged. Without such a precaution, Kemmler's body would be thrown across the room by the force of the current. After each of the eleven straps was tightened, Kemmler tested them, making certain they were properly secured. Those that did not meet his standards were readjusted. "All right, William?" asked the warden. As Durston retightened the rubber skullcap that held a moistened, dense elephant-ear sponge, the caustic soda solution ran down Kemmler's face and neck. Dr. Fell had injected the solution into the sponges with a syringe to allow for better contact with the current and prevent burning or scorching of the flesh.[63] Once finished, the warden stepped back. Like a dog trying to dry himself, Kemmler shook his head. As calmly as before, he said, "Warden, just make that a little tighter. We want everything all right, you know." Warden Durston obliged, once again tightening the skullcap.[64]

All that remained was for Durston to place the leather harness on Kemmler's head. Designed to act as a muzzle, the broad straps went across the condemned man's forehead and chin. Another thinner strap pressed down against Kemmler's nose, flattening it. As this was going on, Dr. Edward Charles Spitzka, who would help perform Kemmler's autopsy, whispered softly, "God bless you, Kemmler." The condemned man, his jaw sealed shut, nodded his thanks.[65]

After securing Kemmler in the chair Durston walked over to Drs. Spitzka and Carlos F. MacDonald, and said: "How long shall I have the current on? You shall say whether it shall be fifteen seconds or three or five."

"Fifteen seconds," Dr. Spitzka abruptly replied.

"That's a long time," said the warden.

"Will you say, Doctor?" said Dr. Spitzka, turning to his colleague. "You have had more to do with these things than I have."

"Well, I have left the matter entirely to you," said the warden. "How much time do you say?"

"Well, say ten seconds at least," replied Dr. MacDonald.

"All right," was the warden's restrained answer, and he turned sharply around and went into the next room.

During his absence Dr. Spitzka asked, "Has any gentleman here a stopwatch?" MacDonald produced one and handed it to Spitzka.[66]

Finally, everything was in order. The dynamo in the machine shop was running at a steady speed and the meter on the wall read a little more than 1,000 volts. His task completed, Warden Durston stepped back from the chair. Turning to the witnesses, the warden asked, "Is all ready?" No one said a word. Kemmler raised his eyes and turned his head just enough to feel the warm sunlight on his face. Then Warden Durston gave the signal. "Good-bye, William." George Irish, a New York state government clerk who "had always been good at that kind of work,"[67] is thought to have pulled the lever. Later at the Osborne House, he bragged about having thrown the switch.[68]

A click was heard and Kemmler's body strained against the leather straps, every muscle in full extension. Kemmler's eyes bulged but did not otherwise move.[69] His body remained rigid except for the right index finger, which contracted, curling so tightly that it dug into the flesh of the first joint, causing blood to trickle onto the arm of the chair.[70]

Next, the condemned man's complexion turned ashen. "Death spots" appeared on his skin. After seventeen seconds, Dr. Spitzka shook his head and declared, "He is dead." Warden Durston gave the signal to stop the flow of electricity. Dr. Alfred P. Southwick, a Buffalo dentist who was an early proponent of the electric chair, solemnly declared the first execution by electricity a grand success, saying, "There is the culmination of ten years' work and study. We live in a higher civilization from this day."[71] The witnesses, who had averted their eyes, now turned back toward the chair. But their sighs of relief turned to gasps of horror as they faced Kemmler's still-twitching body. "Great God! He is alive!" yelled one. "Turn on the current," screamed another. "See, he breathes," hollered a third. "For God's sake, kill him and have it over," shouted a newspaperman, who then fainted.[72] District Attorney Quinby clutched his stomach and ran for the door; once outside, he fainted.[73]

Drs. Spitzka and MacDonald calmly stepped forward to examine the body. Warden Durston began to unscrew the electrode attached to the skullcap. Spittle dripped from Kemmler's mouth as the condemned man continued to breathe, his chest rising and falling convulsively.[74] The medical men, present to pronounce death, listened for a heartbeat and, finding one, signaled Durston to reconnect the electrode on Kemmler's head. Turning toward the warden, Dr. Spitzka shouted, "Have the current turned on again, quick—no delay." Warden Durston ran to the door and sounded the bell twice, which was the signal to the men in the machine shop room to turn the current on again.[75]

Once more the click, and again Kemmler's body, like a toy soldier, snapped to attention. The scene grew more gruesome as the dynamo, now running at top speed, sent 2,000 volts through Kemmler's body. On the other side of the prison, in the dynamo room, convicts were pressed into service, holding the new leather belt on the dynamo; it hadn't been stretched properly before the execution, and it almost fell off several times.[76] Froth oozed out of Kemmler's strapped mouth. The small blood vessels under his skin began to rupture. Blood trickled down his face and arms. Twice Kemmler's body twitched as the current was switched on and off. The awful smell of burning flesh filled the death chamber. Kemmler's body first smoldered and then caught fire.[77]

When the current was finally turned off, Kemmler's body went limp. This time he was dead—there could be no doubt of that. From the moment he first sat down on the chair until the electricity was shut off the second time, eight minutes had elapsed. Kemmler's blackened, smoldering body was left strapped to the chair as the horror-stricken witnesses were marched out of the death chamber into the stone corridors of the prison. A "pungent and sickening odor" followed them. For a long time, no one spoke. Finally, Dr. George E. Fell, the Buffalo professor, turned to a reporter and said: "Well, there is no doubt about one thing. The man never suffered an iota of pain." With the exception of Dr. Fell, few others felt much pride in their participation.[78]

THE AUTOPSY was conducted by Dr. Carlos F. MacDonald, who had been appointed by the governor as medical counsel, Drs. Fell, George

(*New York Herald*, 7 August 1890)

Just prior to Warden Durston's giving the signal: "Good-bye, William"

Shrady, Spitzka, and C. M. Daniels, and New York deputy coroner William T. Jenkins.[79] The doctors laid Kemmler's charred body upon a dissecting table directly in front of the chair and then left the room. Thirty minutes passed before the medical team returned to the execution chamber. Upon examination, Dr. Spitzka found that the corpse was still warm, with a temperature of 98.8 degrees Fahrenheit. Since death was defined medically as the inability of the body to produce heat, Dr. Spitzka decided to delay the autopsy for three hours, hoping to protect the medical team from allegations that Kemmler died at the hand of the "doctor's knife" and not from the executioner's current.[80]

By 9:00 a.m., Kemmler's body had cooled to 97 degrees Fahrenheit. Rigor mortis had set in and the corpse had assumed a sitting posture.[81] As a convict prepared Kemmler's body for autopsy by sponging it down with water, the medical men struggled to straighten the arms and legs. Dr. Daniels inspected the nerve centers—the head, brain, and spinal column—while Dr. Jenkins examined the chest and abdominal cavities. As the internal organs were removed, the two physicians dictated their observations while Dr. Shrady took notes and made drawings. The autopsy was quite extensive, lasting four hours; there was barely a part of Kemmler's body that was not dissected. The heart was normal, except for a slight congestion at the top. The lungs were congested, indicating that the condemned man had suffered from the early stages of pleurisy. The other organs in the abdominal cavity were in a healthy condition. The blood, however, was watery and pale, probably due to the strength of the electrical current.[82] The throat muscles were contracted, and the face frozen expressionless. The eyes remained fixed open, giving the body an eerie appearance.[83]

It was in the head that Dr. Daniels found evidence of the strength of the electrical current's second deadly force. Kemmler's scalp was badly scorched directly under the spot where the electrode was placed on his head. The skull bone was burned dry, and the blood in his head had turned into a blackened powder: disturbing evidence of the terrible punishment that had been meted out on this unsuspecting victim of a scientific experiment.[84]

Dr. Spitzka removed the brain and spinal cord, dividing them up for further study among the participating doctors. To the untrained eye, there was no sign that either had been injured by the electrocution. The internal organs were also divided up among the doctors for microscopic examination in their laboratories. Dr. Fell took several jars home; two contained blood from the right and left sides of Kemmler's heart, one held part of the brain and spinal cord, and one contained some of the skin that was burned by the electrodes. Dr. Daniels took one of the four pieces into which the brain had been divided.[85] Prior to the execution, no adequate preparation had been made to measure the exact amount and duration of the current that flowed through Kemmler. Therefore, despite months of intensive analysis, little useful knowledge was gained about the effects of electricity on the human body.[86]

AT ABOUT 4:00 on the afternoon of August 7, the death certificate was presented to Registrar Grinnell, and a burial permit was granted. Dr. H. E. Allison certified that Kemmler died at 6:43 a.m. on August 6, 1890, from a current of electricity passed through his body. Samuel Miller, a longtime resident of Auburn, was the undertaker. Although Kemmler's brother had expressed his intention to do so, the family never claimed the body.[87]

Kemmler's burial was originally scheduled for 10 p.m. on August 7, but on that night a number of reporters were stationed outside the prison's north carriage gate in anticipation of the burial procession.[88] The principal keeper, Major Boyle, and two assistants had been overheard discussing how they would take Kemmler's body out of the prison yard without being discovered.[89] Now, as the reporters waited, a horse-drawn cart turned down Wall Street and stopped in the shadows.[90] As reporters rushed toward the wooden cart, Boyle abandoned the plan to bury Kemmler that night, delaying it until the following morning.[91]

Finally, at 4:00 a.m., two days after his execution, Kemmler's remains were buried in the prison graveyard in the Fitch Avenue lot adjoining Fort Hill Cemetery.[92] The body was placed in the plain pine coffin built in the prison workshop; it was filled with quicklime and buried in the twenty-first grave in the sixth row.[93] The graveyard itself was in an untended field overrun with wild carrot. The clergymen who in life had devoted so much attention to Kemmler's spiritual well-being neglected to attend his burial.[94]

ALMOST IMMEDIATELY, medical men and the press began to quarrel over who was responsible for the botched execution. The New York Herald placed much of the blame for what it considered to be a "failed" execution on Spitzka: "In a scientific sense Dr. Spitzka may be considered executioner-in-chief. Many of the doctors are not satisfied with the way Dr. Spitzka ran things. They claim he only permitted 1,300 volts when he should have had 1,800, and that many of the horrors of the execution could have been prevented if he had used a current as strong as was intended."[95] Several leading newspapers agreed, quoting Drs. Southwick, Fell, and especially Daniels as blaming Spitzka for having botched the execution.[96] Dr. Southwick said that Dr. Spitzka "got rattled" and ordered the current turned off too quickly. "I am sorry I did not interfere," he said later. Similarly, Dr. Fell said he was "astonished when the current was . . . stopped so soon." Dr. Daniels attacked Spitzka in the harshest terms: "The execution would have been a success had it not been for Dr. Spitzka."[97]

Spitzka reacted with indignation, going so far as to threaten Daniels's life. "If the scoundrel had dared to make any such statements while I was in Auburn, I really believe I would have killed him and had a chance to sit in the fatal chair myself," Spitzka told the New York press. He said it was not his decision to stop the first surge of electricity; by prior arrangement, fifteen seconds was the agreed-upon length of time for the first current. Spitzka said he borrowed the stopwatch at the last minute because no provisions had been made to record the duration of the electric current flowing through Kemmler's body. "I went to Auburn solely to make the examination of the brain," he said, "and not as an assistant executioner."[98]

According to Spitzka, many of the nonmedical witnesses panicked at the muscular convulsions that caused Kemmler's lungs to contract. Spitzka later contended that it was only after the witnesses became distraught that he "made a mistake" and called for the current to be turned on again. He said he had no doubt that Kemmler was dead from the first jolt of electricity. Spitzka attributed Daniels's criticism to the fact that he and Dr. MacDonald supported the appointment of Dr. William Jenkins as head of the autopsy team.[99] Daniels replied to Spitzka's angry letter in one of his own to the editor of the New York Sun, claiming that the so-called interview in the Sun was a complete fabrication. Daniels claimed that the only comment he made concerning Dr. Spitzka was to point out that Spitzka and MacDonald were the official physicians in attendance and that all medical responsibility rested upon them.[100]

Despite his strenuous denials, Spitzka did panic. All the witnesses agreed that while Spitzka was pointing out the postmortem signs of death, Kemmler gasped for breath and Spitzka yelled, "Turn on the current, turn on the current, the man is not dead!"[101]

Dr. Spitzka pronounced the new method of execution a failure.[102] "I have seen hangings that were immeasurably more brutal than this execution, but I've never seen anything so awful," he told reporters. He maintained that if the purpose of adopting this new means of execution was to rid the process of "cruelty" and "barbarity," then electrocution was a dismal failure. Had it not been for the extraordinary courage and uncommon poise of William Kemmler, Spitzka continued, the execution would have been far more ghastly. He added that Kemmler was as "self-possessed as though about to sit in a barber's chair."[103] In extensive comments to the *New York Tribune*, Spitzka elaborated, "All this goes to show that the new method will not take from capital punishment the barbarous features of an execution." Spitzka went on to allege that the Westinghouse Company, which had tried to stop the use of its dynamos, wanted to see Kemmler's execution botched. He charged that they deliberately failed to inform the warden that the Westinghouse dynamos were unsuitable for use in an execution. According to Spitzka, a failed execution would best serve the interest of the Westinghouse Company. It would cause the public to demand an end to electrical executions, and thus disassociate the Westinghouse name from state-sanctioned killing.[104]

As might be expected from a major proponent of the chair, Dr. Southwick was highly enthusiastic about Kemmler's execution.[105] Overflowing with confidence in the future of executions by electricity, Southwick predicted, "There will be hundreds more executions by electricity, for the experiment of yesterday morning was a success. I don't care what anybody says, science has proved that Kemmler died an absolutely painless death." His enthusiasm for electrocution was so strong that he boldly declared, "Ladies could have been in that room and not known what was going on, so silent was the process; not a cry from the subject, not a sound."[106]

Few of the medical men in attendance were as enthusiastic. Dr. Carlos MacDonald, president of the state lunacy commission and a member of the autopsy team, was of two minds. After the execution he told reporters, "I don't think that the execution was as successful as it should have been, because the dynamo was too far away from the death chamber." Nevertheless, he was apparently satisfied that electrocution was superior to hanging. Dr. MacDonald acknowledged that it was impossible to determine whether Kemmler was dead after the first jolt of electricity, but said he was positive that the man felt no pain. He chalked up the problems with the procedure to the fact that it was the first, and blamed its faults on the electricians and mechanics who operated the execution apparatus. "Those in charge were too nervous," he added.[107]

(National Police Gazette, 23 August 1890)

Kemmler entering
the death chamber

Removing his coat

(National Police Gazette, 23 August 1890)

In an interview with the *Auburn Daily Advertiser*, Dr. MacDonald elaborated on what he thought of those in charge: "This is not a case which calls for personalities. Mr. Davis was in charge, with assistants, but I did not see anyone in there in a convict uniform. . . . The test lamps were burning dimly and the current was not up to the power it exhibited when I made the tests in . . . Sing Sing, Dannemora and Auburn. I called Mr. Davis' attention to the fact, and he said at once, 'Yes, there must be something wrong with the machinery.' . . . The current was not strong enough to make the lamps burn at full power. . . . When I started to go into the switch room again, before the execution, I was refused admission. There is no doubt in the world that some of the witnesses should have seen the voltmeter and ammeter register." MacDonald concluded, "If the current had been kept on Kemmler from three to five minutes longer there would have been no trouble."[108]

Dr. George Fell, known for his respiratory apparatus that had succeeded in resuscitating half-drowned and half-suffocated people, also had mixed feelings about Kemmler's execution.[109] "I am still a defender of electrical execution," he said, "but I do not entirely defend the manner in which Kemmler was executed." Kemmler was supposed to die "quickly and painlessly," not suffer as the subject of a scientific experiment. Although he believed Kemmler's death was instantaneous, he argued that the initial current should have been left on longer to finish the job properly. He claimed Kemmler's gasps and convulsions were merely muscular movements in reaction to the force of the electrical current.[110]

Since Kemmler's body was still warm an hour after his execution, Dr. Shrady was convinced that Kemmler did not die instantaneously. There had been speculation that Dr. Fell might try his patented resuscitating machine on Kemmler.[111] Shrady charged that Kemmler could have been resuscitated by Fell's machine, and stated firmly that Kemmler had suffered a prolonged and torturous death.[112] Compelled to explain why he did not use his celebrated respiratory apparatus to revive Kemmler, Fell declared, "I do not believe there was any possibility of reviving Kemmler after the first shock."[113] This was an odd query since the object was to kill Kemmler, not to bring him back to life.

New York deputy coroner Dr. William T. Jenkins was visibly shaken by Kemmler's execution. He condemned death by electricity, saying, "I would rather see ten hangings than one such execution as this." He charged that the execution apparatus was defective due to carelessness, but that even a perfectly functioning mechanism would not have guaranteed success. Dr. Jenkins was particularly upset by the death pallor on Kemmler's face and the froth oozing from his mouth.[114]

In a *Medical Record* editorial, Dr. Shrady speculated that the electric chair might lead to the abolition of capital punishment. Although he agreed that Kemmler's death was quick and painless, he argued that electrocution was far inferior to hanging.[115] He also noted that the execution apparatus was extremely complex, requiring elaborate and careful preparation, increasing the cost and exposing the workmen, witnesses, and even the executioner to potential danger. In comparison to hanging, Shrady maintained, the elaborate preparation required for electrical execution inflicted untold agony on the condemned.[116]

Shrady wrote that science may have triumphed at the expense of humanity. He pointed out that previously the goal of science was to save lives. Anticipating the concerns that medical and scientific professionals still harbor today, he believed that with their involvement in this new form of execution, men of science had tragically served "the whims of a few cranks" who believed they could make murder a scientific achievement.[117]

Erie County sheriff Oliver A. Jenkins, whose official duty it was to attend the execution, was horrified, and openly challenged the opinion of the medical witnesses. The "smell of burning flesh still haunts me," he told reporters. He maintained that Kemmler's execution was a brutal scene. He did not see how Dr. Southwick could characterize the execution as a "grand success." To the lay witness, Jenkins continued, it seemed that Kemmler was coming back to life after the first electrical shock, and that he was in unspeakable pain. Sheriff Jenkins maintained that if Kemmler was actually dead when Dr. Spitzka claimed he was, then electrocution would be a humane method of putting a man to death.[118]

According to the sheriff, since the examination of the internal organs took place after the second jolt of electrical current, the doctors could not know for certain when Kemmler died. Sheriff Jenkins described the scene dramatically; he told reporters he was anxious to leave the execution chamber. "We wanted to go out of that awful smelling place, but we were detained with that body lying broken in the chair, its eyes half open as if sleepily watching us till we had signed our names and certified that he was dead."[119]

As expected, Charles S. Hatch, Kemmler's attorney, denounced electrocution as inhumane. He expressed concern over the regulation of the electric current; he claimed that adjusting the electrical jolt so that it will kill, yet not torture and burn, its unfortunate victims was an impossible task. He noted that a hangman could rely on many years of experience with the rope; in contrast, "[With] the use of electricity, those who employed it were obliged to rely upon a dynamo generating a force that they could not fully understand or control." He believed the tragic saga of Kemmler's death guaranteed that the legislature would repeal the law mandating executions by electricity. He predicted that the governor would ban any future scientific experimentation with human lives. "The position that I, as counsel in the case," took had been proven, he asserted.[120]

Charles R. Huntley, the manager of the Brush Electric Light Company, who attended the execution as a private citizen, agreed with Kemmler's lawyer. The *Buffalo Evening News* quoted him as saying, "It was one of the most horrifying sights I ever witnessed or expect to witness. There is no money that would tempt me to go through the experience again. I will see that bound figure and hear those sounds until my dying day."[121] As an electrician who was proud to be associated with an industry that did much to improve men's lives, Huntley told the *Buffalo Courier* that he could not abide by "the prostitution of that industry."[122]

According to Huntley, even if electricity were quicker and less painful than hanging, it inflicted much more mental anguish on the victim. The preparation necessary for electrocution had to be a "fearful torture," he told reporters, whereas putting the noose around the neck of the condemned man took only a moment. The man who is executed by electricity must have "his head shaved, the electrode must be firmly fixed at his spine and on his head, and he must be strapped to the chair," Huntley said. "The agony to a man less stoical and indifferent [than Kemmler] must be inhuman and unbearable . . ." he continued.[123] To allow for the passage of a significant quantity of electricity, he added, the executioners should have used a thicker wire; they used a number 12 wire, no thicker than the lead of a pencil, whereas a number 6 wire was the size of a telegraph wire and would have been better suited to the purpose.[124]

Despite Warden Charles Durston's initial apprehension, and perhaps even contrary to his true opinion, he claimed that the execution was "perfectly satisfactory." While acknowledging that he had never witnessed a hanging, he was quite sure that electrocution was "infinitely superior" to it. Durston's comments in praise of the execution differed significantly from those of the reporters whom he had unlawfully invited to attend the execution. Perhaps Durston deliberately violated the execution law's ban on reporters, so that the truth could be told—but not by him.[125]

Likewise, Governor Hill remained steadfast in his support for electro-cution. His first message to the legislature as chief executive of the state of New York had proclaimed his belief in electrocution, and he did not waver from his original stance, even after Kemmler's death.[126] In a prophetic summation of an argument that has endured to the present day, Hill asserted that killing a man could hardly be viewed as humane, no matter what method was employed. Hill was "besieged almost as closely as Warden Durston" by reporters following the execution. Finally, after extensive questioning about the electrocution, Hill, growing increasingly irritable, announced that he would answer no further queries about Kemmler and his execution.[127]

IGNORING THE GAG ORDER mandated by the new electrical execution law, the reaction of the New York press was quick and decisive; most took a negative view.[128] Kemmler's death proved not only to be a watershed event for capital punishment, but it also had a significant impact on the ever-intensifying struggle between newspapers that practiced information journalism and those that engaged in storytelling. The controversy sur-rounding the Kemmler execution provided an opportunity for the New York press to battle over their contrasting journalistic styles.[129]

The highly popular *New York World*, with a circulation of 600,000 and a well-deserved reputation for sensationalism, denounced the execution as "very cruel and very shocking." From all visible indications, the paper wrote, Kemmler died in slow, torturous agony. His hideous death sick-ened the witnesses, and its effect on the public was still more shocking.[130] The *New York Press* agreed, arguing that the people of New York State did not want to torture anyone, "even the most desperate criminals," nor did they enjoy "the smell of burning human flesh."[131] The *Troy Press* added that the grisly torture of a condemned prisoner had tarnished the image of the great state of New York. It claimed, in a sensational historical compar-ison, that the electrical contraption used to kill Kemmler was reminiscent of the elaborate torture chambers of the Dark Ages.[132]

Like about twenty other New York papers, the *World* called for a repeal of the electrical execution law, saying that "the first experiment in electrical execution should be the last." The *World* added that the law was especially reprehensible because it permitted the authorities to perform a "judicial killing by torture in secret." The paper further pre-dicted that as long as the electrical execution law was on the books, juries

(*Electrical World*, 16 August 1890)

The switchboard, containing a voltmeter, a resistance box, incandescent lamps, an ammeter, a regulating switch, and the lever

would fail to convict in capital cases, unwilling to condemn a man to death by torture.[133]

In direct ideological opposition to the *World* and other papers that had entertainment as a principal feature were the *New York Times* and the *New York Tribune*. Both the *Times* and the *Tribune* tended to take a calm, measured approach that reflected the intellectual orientation of many of their well-to-do, highly educated readers.[134] Respectability and "decency" were the working philosophies of these papers, which saw their primary goal as the presentation of rational, verifiable information. Both the *Times* and the *Tribune* were enthusiastic supporters of the electrical execution law. Unfortunately, the *Times* lamented, the first electrical execution was "badly bungled." It feared that legislators might back away from this scientifically advanced, more humane method of execution. The paper reminded its readers that the initial intention of the electrical execution law was to devise a more merciful method of putting condemned murderers to death. There was every reason to believe, the *Times* held, that electrical execution would be quick and painless once administered correctly.[135]

In keeping with its conservative orientation, the *Times* maintained that reports of the execution were exaggerated and manipulated by those who were determined to make the electrocution appear a failure. It contended that the execution seemed more brutal than it actually was because of the novelty of the process and the mystery connected with the effects of electricity on the human body. Departing from its self-proclaimed emphasis on facts, the *Times* claimed that unnamed individuals wished to protect the commercial interests of the Westinghouse Company by discouraging further electrocutions. It also insinuated that the last-minute changes by Warden Durston might have permitted some tampering with the execution apparatus by Westinghouse operatives.[136]

Similarly, the *Tribune* argued that the Kemmler execution should not be considered a failure. It claimed that hanging's "utterly brutal" results were tolerated only because society was accustomed to the practice. Anyone who had ever witnessed a hanging could testify, the paper maintained, to the "unspeakable horror" suffered by a hanged man. While conceding that the spectacle of Kemmler's execution was equally dreadful to those who witnessed it, the *Tribune* argued that the condemned man died instantly, thus experiencing no pain. It also pointed out that this initial effort was merely an experiment, and future electrical executions were certain to be more successful.[137]

The *Tribune* criticized Warden Durston for failing to install properly the electrocution machinery and abruptly changing locations shortly before the execution. The paper faulted the Westinghouse dynamos for not maintaining a steady or strong current. It also said the length of time and the number of volts required to extinguish a human life should have been decided upon long before Kemmler was strapped to the chair. A public outcry against the new law was to be expected, the *Tribune* continued, noting that unscrupulous elements in society would undoubtedly attempt to exploit Kemmler's execution by demanding the abolition of capital punishment. In the interest of "mercy and civilization," the *Tribune* argued, the electrocution law should remain in effect, at least until further scientific experiments could be conducted.[138]

Smaller papers, relatively unencumbered by national reputation or political alliances, grappled with the major issues surrounding Kemmler's death. These papers tended to be less concerned with matters of a political nature, and more interested in the immediate ethical questions raised by the execution. The *Buffalo Evening News* maintained that the doctors and scientists had no moral right to experiment, even upon a "brutal murderer."[139] The *Buffalo Express* concurred, claiming that while Kemmler deserved to die, the shocking circumstances surrounding his death assured that no other "poor wretches will be made the sport of scientists and mock humanitarians."[140] In calling for a return to hanging, the *New York Sun* asserted that "scientific curiosity ha[d] been gratified" by the awful experiment performed on William Kemmler, and assuredly other criminals would not be made to suffer similar fates. The *Sun* added that the good citizens of New York would no longer tolerate further tests of "this new experimental science of man—killing by electricity."[141]

Besides the major wire services, the *New York Herald* was the only newspaper that managed to place a reporter at the execution. The scribe poignantly noted that "Kemmler paid double penance to the law . . . a penance for his crime and a penance for his childlike trust in men who by their carelessness have brought shame upon the great state whose servants they are." He placed substantial blame for what he called a "grievous failure" on faulty instrumentation, claiming, "The instruments were stolen in the first place. They were admittedly imperfect." Acknowledging that Westinghouse, under pressure from critics, had offered only reluctantly to build a dynamo designed specifically for executions, the scribe nonetheless contended that "the events of today prove either that the dynamos were faulty or that the interested [Westinghouse] company had bribed someone to make them so."[142]

Several newspapers placed the blame on the medical experts for botching the execution. The *Buffalo Evening News* condemned the doctors for proceeding with the execution despite their limited knowledge of electricity's effect on the human body. The *News* asked whether the experts were certain life was extinguished after the first shock. It seemed to the *News* that Kemmler's labored breathing and the spittle that flew across the room from his contorted lips were signs of unbearable pain. "Why was it necessary to cover all but the victim's lips [the only part visible from his leather mask] and bind him hand and foot if death was to be painless," the paper demanded.[143]

Although the big-city *Times* and *Tribune* criticized the handling of the execution, they remained in favor of retaining the electrical execution law. Many of the smaller metropolitan papers, however, categorically denounced both the law and the electrocution. The *Buffalo Evening News* observed that the doctors claimed Kemmler was dead after the first shock of seventeen seconds. But, the paper asked, why did they need to turn the current on again if Kemmler had already died, subjecting his body to four and a half minutes of unnecessary abuse?[144] The *Buffalo Express* opined that the experts still did not have a thorough understanding of electricity. The *Express* believed electricity should not even be used for commercial purposes, and certainly not as an agent of judicial killing. The *Express* argued that strangulation was the least painful method of execution. "Half-hanged and half-drowned men," it continued, "have afterwards described its sensations as positively pleasurable."[145]

While acknowledging that this was the first execution of its kind, the *Buffalo Courier* contended that it did not believe an electrical execution could ever be more humane than a skillfully conducted hanging. The electrical apparatus was "vastly more complicated"; thus, its reliability could never be assured. From Benjamin Franklin to Dr. Alfred P. Southwick, advocates thought it would be the most humane method science could ever devise. Unfortunately, the *Courier* maintained, the facts did not bear out the scientists' theories. Replacing hanging with electrocution only introduced new elements of terror and uncertainty.[146]

The ironic possibility of martyrdom for electrocuted felons was explored on numerous editorial pages. The *Buffalo Express* was concerned that the Electrical Execution Act would cause juries not to convict guilty murderers, and that the public might turn against capital punishment. If life imprisonment was substituted for the penalty of death, the *Express* reasoned, prisoners may be pardoned by a "soft-hearted and soft-headed" governor, or released by a partisan judge.[147] Although it was in the minority, the *Buffalo Evening News* disagreed, contending that executions, by their nature, were cruel and barbaric, and that life imprisonment was the only just and humane punishment for murder. It concluded that the execution of William Kemmler had energized the anti–capital punishment movement. If a bill to abolish capital punishment became law during the next legislative session, the *News* continued, it should surprise no one.[148]

The *New York Press* added to its charge that electrical execution failed to meet its humanitarian goals by further claiming that the new method might actually be counterproductive, that is, it might promote murder.

Ten times more sensational than a hanging, the *Press* continued, electrocution would make "silly women send ten times as many flowers . . . to bloody-handed brutes as they do now." Finally, the *Press* charged that the spectacle of electrocution would turn "the worst of ruffians into a hero and a martyr." It confidently concluded that "the age of burning at the stake is passed; let the age of burning at the wire pass also."[149]

Local papers were the farthest removed from the big-city battle over circulation and journalistic philosophy but were closest, both in proximity and sentiment, to the actual event. In their coverage of Kemmler's death these newspapers echoed their relatively larger counterparts like the *News* and the *Sun* in their incredibly detailed coverage of the story, but they sided with the *Times* and the *Tribune* in their position regarding the means of execution. The *Auburn Daily Advertiser* and the *Utica Herald* were two small upstate newspapers that endorsed electrocution unconditionally. The *Advertiser* complained that the graphic reports of the brutality of Kemmler's execution had been greatly exaggerated. It quoted an unnamed witness who declared the execution a success, claimed it was far superior to the old-fashioned method of hanging, and maintained that Kemmler died quickly and painlessly.[150]

(*National Police Gazette*, 23 August 1890)

Warden Charles Durston

Reminiscent of the *Times*'s conjecture, the *Advertiser* claimed that Kemmler's execution was merely an experiment, and those who claimed it was a failure must have been "very friendly to the Westinghouse system of lighting." The paper went on to suggest that the Westinghouse Company was aware that alternating current would probably burn the body, but did not volunteer this information in the hopes that the execution would be bungled and the law repealed.[151]

In a display of hometown pride, The *Auburn Daily Advertiser* defended Warden Durston's handling of the execution, alleging that the state supplied him with faulty equipment and failed to furnish adequate professional support. It maintained that the scientists present at the execution had insufficient knowledge of the effect of electricity on the human body and could only guess how long the current should have been left on. The paper claimed that Kemmler's muscle contractions were to be expected and said they should not be taken as evidence of suffering. From the time Kemmler left his cell to the time he was declared dead, the *Advertiser* noted, less than eight minutes had elapsed. In a typical hanging, the murderer was left dangling anywhere from fifteen to thirty minutes before being pronounced dead. The paper closed with the prediction that in time, the skeptics would come to support execution by electricity as a humane alternative to hanging.[152]

The *Utica Herald* declared electrical execution a success, saying, "A better way than by hanging has been found to dispose of capital criminals." It reported that all the medical gentlemen present agreed that death was instantaneous. Many of the reporters, the *Herald* claimed, were prejudiced against the electrical execution law; they ignored the facts and published "largely imaginative" reports. The paper approvingly quoted witness Dr. Batch, executive director of the State Board of Health, as saying he had favored hanging until he witnessed the quickness and certainty of Kemmler's death.[153]

As the Kemmler controversy reached across the Atlantic, the London papers united in a condemnation of the new American method of execution.[154] The *Times* of London said that although the motive behind the electrical execution law was admirable, it was impossible to imagine a "more revolting exhibition."[155] Kemmler suffered greatly, the *Times* declared, much more than he would have under the old method of execution. The paper argued that electricity was an unreliable and uncertain execution method, noting that many people who are struck by lightning do not die. The London *Standard* agreed, pronouncing Kemmler's execution "a disgrace to humanity," and predicted that it would have a demoralizing effect on the American people. "Horrible and atrocious" was the London *Telegraph*'s characterization of the execution. And the London *Chronicle* stated bluntly, "The guillotine and the hangman's rope do not compare in cold-blooded barbarism with this achievement of modern science."[156]

The *Times* of London observed that despite their abhorrence for Kemmler's execution, none of the London papers suggested that capital punishment should be abolished. Indeed, the *Times* opined that the challenge facing America was to find a "decent and private" way of putting murderers to death. The execution of criminals was a "sad necessity," the *Times* maintained, but added that it was a duty that must be undertaken with as little "parade" as possible. It noted that a "civilized" nation may "reasonably be ashamed" of its murderers, but the more "quickly and quietly" they were dispatched, the more likely the execution would deter others.[157]

LESS DISPASSIONATE were the opinions of Kemmler's few friends and those close to the case. Westinghouse attorney Paul D. Cravath said that as expected, Kemmler's execution was a failure. "As a humanitarian I hope, as it has now been proven, that killing a man by electricity is the height of cruelty." Cravath went on to predict that some less brutal execution method would be found. Electricity was not dependable because the meters used to measure it could easily malfunction, he continued. In addition, the execution apparatus was made up of "delicate" machinery that could easily malfunction. Therefore, the executioner could never know precisely the force of the electric current or the effect it might have on the man strapped to the chair.[158]

The secretary of the Westinghouse Company, A. T. Howand, condemned Kemmler's execution and labeled it "cruel and against all moral and civil laws." He hoped this would be the last electrocution, adding sarcastically, "I am inclined to think Kemmler was not killed by electricity, but that when the application of the 1,300 volts roasted and sizzled him the executioners completed the horrible work by hitting him on the head with a club." Becoming serious again, he claimed the people of America were not "barbarians" and would no longer support the horrible spectacle of electrical executions.[159]

Breaking his long silence, George Westinghouse denounced the execution as a savage affair. "They could have done better with an ax," he said, adding that his predictions that the electrocution would be a complete failure had come true. In an apparent reference to the Edison Company's role in helping Harold Brown obtain Westinghouse dynamos, Westinghouse said, "The public will lay the blame where it belongs, and it will not lay on us." Westinghouse had always claimed that with the single exception of sending witnesses to testify in the evidentiary hearings on whether electrical execution was cruel and unusual punishment, his company had not tried to impede Kemmler's execution.[160]

Soon after he made that statement, however, Westinghouse made public a correspondence between Roger Sherman and himself that revealed that Westinghouse had made extensive efforts on Kemmler's behalf. In the letter, dated May 7, 1890, Westinghouse told the lawyer how to make the case against the use of Westinghouse alternating-current generators. He suggested that the dynamos Harold Brown had supplied the state were not appropriate for their intended use. He noted that the generator required the use of a boiler, steam engine, piping, wires, belts, and many switches and electrical connections, all of which could malfunction at an important juncture. Furthermore, he claimed, the generators would have to run at an abnormal and dangerous speed to obtain the pressure required for the execution, increasing their chance of failure.[161]

Westinghouse also suggested that a storage battery with an induction coil and a circuit changer would do nicely, and there was no reason for a more elaborate apparatus. Since the storage battery would only be able to deliver direct current, it would eliminate any possible use of alternating current. In his final remarks, Westinghouse charged that Edison had tried to gain a commercial advantage by securing the use of Westinghouse equipment for Kemmler's execution. This would allow Edison to brand alternating current as too dangerous for commercial or residential use.[162]

Edison and his associates took a markedly different view. The treasurer of the Edison General Electric Company, F. S. Hastings, told the New York Tribune he was certain that Kemmler had died a swift and painless death. Kemmler's grotesque facial distortions and bodily movements following the first shock were merely muscle contractions, Hastings argued, and did not indicate any suffering. He noted that in experiments at Edison's laboratory, burning of the skin only occurred when the current was left on long enough to dry the sponges on the electrodes, and said this was probably what happened in Kemmler's case.[163]

THOMAS A. EDISON laid the blame for botching Kemmler's execution on the medical doctors. Ever the pragmatic inventor, Edison attacked the doctors as acting merely upon theory and ignoring the practical realities of electricity. He explained that from a theoretical point of view, the doctors were right in attaching the electrode to the base of the skull, since it was the nerve center of the human body. In practice, though, Edison claimed they were wrong.[164]

Edison further argued that most accidental electrocutions had occurred when a person unknowingly touched a live wire with his hands. Therefore, the hands should have been immersed in a solution of "caustic soda" to afford a reliable contact. He pointed out that electricity flows much more freely through body fluids than it does through the thick bones of the skull.[165] The blood in the hands and arms was a good conductor of electricity and would have been a more effective point of contact. In addition, the hair of the head, as a nonconductor, offers resistance to electrical currents.[166] As proof that Kemmler received only a fraction of the electrical charge, Edison pointed out that Kemmler's skin was merely burned at the points of contact. "Had he received the full 1,300-volt current for the length of time stated," Edison said, "he would have been carbonized or mummified."[167]

When asked how he accounted for the sporadic muscular movement and the loud uncontrolled gasping that followed the first jolt of electrical current, Edison said he had "no doubt" that Kemmler was dead. He attributed the reactions of Kemmler's body to the muscle reflexes that often accompany traumatic death. He also questioned the use of the lamps that were attached to the circuit in order to measure the voltage: "If the report that 20 incandescent lamps were in the circuit and illuminated at the time the current was applied to Kemmler is true the explanation is in part at least furnished. These lamps offer resistance to the flow of the current, and 20 of them would greatly reduce the potential of the current. They should have been removed from the current." He closed by saying that now that the doctors had gained some practical experience, he was confident that the next electrocution would proceed much more smoothly.[168]

Despite Edison's authoritative criticism, Dr. Fell continued to stand by the method used at Auburn, claiming that it was superior because the current's close proximity to the brain immediately destroyed the nerve

(National Police Gazette, 23 August 1890)

Warden Durston reads Kemmler the death warrant.

centers and consciousness of the criminal. He insisted that Edison's method would not reliably obliterate all feeling as quickly as electrodes placed at the head and spine.[169]

THE RANCOROUS DEBATE about Kemmler's execution left many unanswered questions and lingering doubts about the science, humanity, and success of the first electrocution. The avarice surrounding the ordeal served not only to seal inexorably the horrible fate of one man, but it created a technology of killing that continues essentially unchanged to this day. But how was it that William Kemmler became the first person to die in the electric chair? And, more important, how was it that Thomas Edison became directly involved in the invention of the electric chair? To begin to answer these questions, we must return to the early history of electricity and to the struggle between Edison and Westinghouse for control of the emerging electrical power industry: what is generally referred to as the Battle of the Currents.

CHAPTER TWO

The Battle of the Currents

Edison versus Westinghouse,
DC versus AC

I N THE 1870S, most private residences and commercial buildings were illuminated by gas. Progress was swift in coming, however, and electrical arc lighting soon made minor inroads into the gas industry. First discovered in 1809 by English scientist Sir Humphry Davy, arc lighting is produced when an electrical current jumps back and forth from one carbon filament to another, creating a zigzag pattern reminiscent of the electrical charges that brought life to Dr. Frankenstein's creation.[1]

Most electrical inventors dedicated themselves to improving the arc light. A distribution system was already in commercial operation, and recent innovations in electrical generation and illumination made the future of arc lighting appear bright.[2] Although city streets were still lighted by gas, arc lighting was clearly superior for this purpose. Because it produced a brighter and more intense light than gas, arc lighting did a better job of illuminating large exterior spaces such as city streets and parks. Some commentators even predicted that it might help to curb street crime.

Street lighting, however, was really its only practical application. Unlike a standard gas burner with its warm, soft glow, an arc light was distinguished by a bright, flickering glare that was difficult on the eyes.[3] As the arc burned, the distance between the carbon filaments had to be adjusted, requiring constant attention. The resulting smoke and smell meant that arc lighting could not be used indoors, especially in small rooms.

(Edison National Historic Site)

Thomas Edison

Despite these drawbacks, inventors and scientists searched continuously for ways to make arc lighting more practicable. Thomas Edison believed this work was futile. From his study of the gas industry he knew that residential and commercial users accounted for 90 percent of the lighting market, and he doubted that arc lighting could ever be made practical for home or office use. Edison understood that while arc lighting had its advantages, the average consumer needed to be able to operate each lamp independent of the others around it—a major feature of the gas burner—and one that arc lighting could not provide. To compete with gas, it was necessary to invent a light that had as good a glow as gas, but did not consume oxygen, soil adjacent walls, or pose a fire hazard. Edison's aim was to design an electrical lighting system modeled after gas—one that was as cheap, convenient, and practical as gas—only better.[4]

Contrary to popular myth, Edison did not actually invent the lightbulb. By the time he began his quest for incandescence, work on the modern lightbulb had already begun. In 1820 De La Rue produced an incandescent lamp by passing an electrical current through a coil of platinum wire sealed in a glass vacuum tube.[5] Then, in 1850, Edward Shepard made a lamp out of a weighted charcoal cylinder pressed against a charcoal cone. The lamp could not maintain a sufficient vacuum and, thus, was able to burn for only a short time.[6]

In the summer of 1859, Moses G. Farmer, a professor at the Naval Training Station in Newport, Rhode Island, lighted the parlor of his Salem, Massachusetts, home for several months. A decade later, the Russian Lodyguine illuminated the dockyards at St. Petersburg with two hundred incandescent lamps.[7] Still, though, after forty years of experimentation, no one had been able to produce a practical incandescent lamp that would glow for more than twelve hours; the fragile filament kept breaking.

Unlike the others, Thomas Edison approached the search for a dependable incandescent lamp filament with commercial considerations in mind. In order to employ a conducting wire of a reasonable size, Edison figured that the electrical force could be no more than 110 volts.[8] A lower voltage would make the development of an incandescent light much easier, but the corresponding increase in the size of the copper conducting wire would make its cost prohibitive. From painstaking experiments, Edison had learned that a suitable filament material must have both a high resistance and a small radiating surface. He first tried paper carbon, then wood carbon, and even carbonized broom corn. All were easily brought to incandescence, but none would burn for more than an hour or two.[9]

Even if a suitable filament material could be found, Edison knew that the success of his invention was contingent on appropriate manufacturing and distribution. Although he was a dreamer, a visionary, and an eternal optimist, he was nonetheless aware of the realities of the marketplace, and he understood that his ambitious plan would require ample financial resources. His elaborately equipped laboratory already had a highly trained staff, but the magnitude of his expectations demanded more. Before beginning work on the incandescent lightbulb, he needed to secure financial backing. His earlier experiences with the telegraph and the telephone reminded him that the commercial success of any invention often depended on adequate financing and aggressive marketing.[10]

To that end, on October 15, 1878, Thomas Edison and a dozen New York City businessmen incorporated the Edison Electric Light Company with the backing of Drexel, Morgan, and Company, the leading Wall

Street investment-banking firm. The company had no trouble raising $300,000 in capital stock.[11] The response to the initial offering was a tribute to Edison—not merely to his reputation of being an inventor who could turn a profit, but to the perception that he was a man who could succeed gloriously where others had failed.

Edison set to work, experimenting with different materials and designs in search of a practical incandescent lightbulb. He undertook the task in his usual manner, employing a dogged method of trial and error. Later, Nikola Tesla, the brilliant Croatian-born Serbian scientist and a former Edison employee, would speak with amusement of Edison's method. "If Edison had to find a needle in a haystack, he would proceed at once with the diligence of the bee to examine straw after straw until he found the object of his search."[12] Tesla, like many others who worked with Edison, thought that if Edison had only been open to using scientific theory, he would have saved 90 percent of his labor.

Edison's vast laboratory in Menlo Park, New Jersey, provided him with many "haystacks" through which to search. He kept the lab fully stocked with everything from the exotic to the ordinary, from an elephant hide to the hair of a native Amazonian. A colleague once jokingly remarked that Edison's storeroom contained everything, including "the eyeballs of a US senator."[13] Indeed, Edison used lampblack from the glass chimneys of his laboratory's kerosene lamps to make a powder. Once rolled into a thin thread, these carbon deposits made a perfect filament.[14]

The real breakthrough came in the autumn of 1879. Edison had just obtained a new Sprengel air pump that was capable of producing an almost perfect vacuum inside a glass bulb of one hundred-thousandth of an atmosphere. Edison readily modified the pump to produce one millionth of an atmosphere. The increased vacuum enhanced the durability of the filament, now shaped like a hairpin.[15] Edison's first lightbulb lasted only a few hours, but an often-told story depicts the inventor exclaiming "Eureka!" after a second lightbulb purportedly burned for over forty hours. "I think I've got it," declared the proud Edison. "If it can burn forty hours, I can make it burn a hundred."[16]

This account is, by all evidence, fiction. The famous second lightbulb actually burned for only fifteen and a half hours, but the team of scientists who witnessed the event got caught up in the excitement of the discovery and embellished the story a bit. Their embellishment was not uncharacteristic, however; Edison's vibrant, infectious enthusiasm over even the smallest discovery led him to many "Eureka!" moments, most of which never amounted to much. One anecdote tells of Edison, on his daily walk around the grounds of his laboratory, being so amazed upon finding a certain bug that emitted a strange odor that he wrote to Charles Darwin about it.[17]

Even with the incandescent light perfected, there was still much work to be done. The new pear-shaped lightbulbs operated in parallel circuits rather than the series circuits of the pencil-shaped carbon rods used in arc lighting. While arc lighting used a constant current dynamo and adjusted its voltage to the number of lights, the new Edison dynamo had to be constant voltage in order to vary its current output to compensate for the number of parallel circuits.[18] This was not an easy task. It would take years of painstaking research.

In a little over a year, Edison ran out of money. His exhaustive experimentation and extravagantly elaborate laboratory were draining financial resources, and although his backers knew of his early success with the incandescent lamp, they remained skeptical of its commercial value. In a private meeting that must have been humiliating for the proud inventor, Edison pleaded for further support. He desperately needed money to construct an experimental power station at Menlo Park, but his board of directors were hesitant to go along. One even wondered aloud if the incandescent lamp was anything more than "a laboratory toy."[19]

Fortunately for Edison, rumors of his discovery began to leak out of his laboratory. Passengers on the Pennsylvania Railroad traveling between New York and Philadelphia told of bright lights emanating from the Edison laboratory and its adjacent buildings. The "buzz" that surrounded Edison captured the attention of investors, who scrambled to purchase the few publicly traded shares. At one point, Edison stock rose to an astounding $3,500 a share.[20] Nonetheless, Edison's main backers remained reluctant to invest more money. Several opted to sell their shares and reap the profits.

Edison was confident in the economic potential of the incandescent lightbulb, but he realized that his reputation alone could not carry him through this exceedingly expensive endeavor. He decided that some favorable publicity was needed to persuade his investors. An article strategically placed in the *New York Herald* did the trick. The author, Marshall Fox, with the help of Edison and his staff, described and illustrated the new Edison system of incandescent lighting. The story detailed Edison's discovery of how to produce light from a "tiny strip of paper." Waxing poetic, Fox described the incandescent lamp as giving off "a bright, beautiful light, like a mellow sunset in an Italian Autumn."[21] The story was widely read, and several rival inventors used the details published in the *Herald* to help them catch up with Edison.

A month later at Menlo Park, Edison held the first public demonstration of incandescent lighting. On New Year's Eve 1879, over three thousand visitors descended on the Edison laboratories. Edison told the expectant crowd that he intended to light up the surrounding village and countryside with eight hundred incandescent lightbulbs, and then, swept up in the moment, he announced that he would soon be illuminating the great cities of Newark, Philadelphia, Boston, and New York. Edison further stated that his magical new system would be affordable for the common man; his lightbulb would be priced at twenty-five cents and would cost only a penny a day to operate. Get ready, he told them: the age of electricity was upon them.[22]

Not everyone at the Menlo Park demonstration came to celebrate Edison's achievement. In an industry riddled with deceit and thievery, there were some who blamed Edison for their failures. Among the crowd was William E. Sawyer, a drunkard who at one time had been a worthy rival. He claimed, not without some justification, that he was the true inventor of the incandescent lamp. Sawyer had hidden a piece of insulating wire under his coat sleeve, with which he planned to short-circuit the entire system. Edison, wary of security breaches, had insulated each group of lamps with safety fuses, so when Sawyer crossed the wires just four lamps went out. Sawyer was hastily ushered out of the building, all the while shouting invective at Edison.[23] Although a security force was on hand, some equipment was damaged, eight lightbulbs were stolen, and a few visitors who disregarded warning signs ventured into the dynamo room and found their watches magnetized. A well-dressed lady, upon bending down to examine something on the floor near one of the dynamos, had all of her hairpins fly suddenly out of her hair.[24]

As its renown grew, the incandescent lamp became not just a wonder of science, but a status symbol of fashionable society. In 1883, Alva Vanderbilt staged an elaborate ball marking the emergence of the Vanderbilts onto the New York social scene. The gala affair was a conspicuous display of haute couture, and Alva's sister-in-law, Alice, descended her grand Victorian staircase decked out as the newest wonder of the world—the incandescent lightbulb.[25]

Satisfied that Edison's endeavor was of practical benefit, the directors of the Edison Electric Light Company were willing to invest in further research and development. Edison had achieved his goal, but the cost had been considerable. Edison's competitors had gotten a good look at detailed drawings and descriptions from the *Herald* story, and so were supplied with blueprints for their eventual assault on the Edison electric lighting system. In addition, Edison was already under heavy fire from early detractors and skeptics. M. De Moncel, the leading French authority on electric lighting, told the *New York Times* that there was nothing new or novel about Edison's incandescent lamp, that its durability was questionable, and the theories advanced in support of the new lamp violated "the laws of electricity as studied by Ohm and Joule."[26]

The public demonstration at Menlo Park, while impressive, really did not prove the commercial value of incandescent lighting. Some critics, such as the respected electrical engineer Elihu Thomson, publicly stated that Edison's incandescent lamp was impractical and could not compete with arc lighting. According to Thomson and others, Edison's lighting system would require "all the copper in the world."[27] To prove his critics wrong, Edison had to bring his invention into use in actual homes, offices, and factories.

Edison's ability to deliver on his promise was due mainly to his extensive research laboratory at Menlo Park. While other inventors labored alone or with one or two associates, Edison employed a staff of sixty skilled professionals. There were electricians, mechanics, physicists, chemists, mathematicians, and even glassblowers. This gave Edison a versatility and scope possessed by no other inventor of the day. Not only could he work on more than one problem at a time, but he could work on several different inventions at once, thus enhancing productivity, technological progress, and, ultimately, profit.[28] The modern research laboratory is a direct descendant of Edison's Menlo Park facilities. It was Edison who brought invention out of the backyard shed, carriage house, or barn, and into the twentieth century.[29]

BEFORE EDISON could launch the "age of electricity," a central power station was needed. A suitable location was found at 255–257 Pearl Street in lower Manhattan, near the commercial and financial districts of New York City. Edison also bought the old ironworks of shipbuilder John Roach in downtown New York, and this became the nucleus of activity for the hundreds of Edison employees who set to work building generators for lighting the nation's homes, streets, and businesses.[30]

To ensure the success of the Pearl Street power station, Edison conducted a detailed customer survey to determine how gas was used in the area. The findings provided a reliable measure of the probable demand for electricity. This allowed Edison to size his system, and thereby maximize its economic potential.[31] Since most gas users were willing to pay an equivalent price for electricity, Edison priced his system accordingly. To keep costs down, he converted existing gas pipes and burners to carry his electric lines. This is why the early electrical fixtures looked just like the old gas burners.

Both Mrs. William Vanderbilt's and J. P. Morgan's homes were wired this way. The system was far from perfect, however, and one rainy day the wires of Mrs. Vanderbilt's electrical system short-circuited, starting a fire. Upon his swift arrival, Edison ordered the power shut off, and the fire went out immediately. But Mrs. Vanderbilt became "hysterical," and when she learned that there was a steam engine and a boiler installed in her basement, she declared that the entire thing must be taken out: she "could not live over a boiler."[32]

J. P. Morgan, whose home was wired in the autumn of 1882, directed that his steam engine and boiler be located away from his house in a pit under his garden. This worked for Morgan, but his Murray Hill neighbors complained that they could not sleep at night due to all the racket that came from the electrical apparatus, and they repeatedly threatened litigation. Morgan, greatly enjoying his new lights, paid no attention. Even when a short circuit ignited a small fire that damaged the walls and floor of Morgan's library, his enthusiasm for electric lights remained unaffected.[33]

If electric lighting was to replace gas, it would have to contain many of the same features that made gas popular. Because there was grave concern that the high voltage necessary to transmit electric current was dangerous if touched accidentally, Edison insulated all copper conductor wires and placed them underground. He also obtained a written statement from the New York Board of Fire Underwriters declaring that insurance rates would remain unchanged for those buildings converting from gas to electric lighting.[34]

To encourage offices and private residences to switch to electricity, the Edison Electric Light Company offered to pay all installation charges until the customer decided if he wanted the electric light permanently. For several months, no bills were sent out as the station engineers worked out the bugs in the system. After the first three months, each customer was charged a flat rate based on the number of incandescent lamps, regardless of actual electrical usage. Soon after, Edison developed an electrochemical device that crudely but adequately measured monthly energy consumption.[35]

Pearl Street was plagued with problems, however, and its opening was delayed as one snag after another presented itself. Edison was steadfast in his concern for the safety of his customers, and insisted on running the electrical wires under the ground, despite pressure from his closest associates and city hall, where it was considered an unnecessary and expensive precaution. One city official remarked, "Some electricians wanted all the air, Edison only asked for the earth."[36] Nonetheless, Edison prevailed, and in April 1881, the city of New York required that all electrical wires be placed under its streets,[37] a decision that, to this day, has improved the appearance, if not the safety, of New York City's streets.

Placing wires underground required months of digging, measuring, and adjusting. True to Edison's style, these working days typically dragged on well into the nights. The head of the generator factory, John Kreusi, recalled a night when, after an exhausting day, he, Edison, and Edison's right-hand man, Charles Batchelor, all slept in an abandoned shop where iron tubes were stored: "There was room on the floor for one, on a work bench for the other. They drew lots and Edison drew the shortest; so he was reduced to lying down for the night on the iron tubes. I remember that he had on a very light colored suit which, by morning, was marked with streaks of tar from shoulders to feet, for the warmth of his body had softened the tar."[38]

The work progressed, and soon Pearl Street was outfitted with all the equipment needed to power New York's financial district, complete with the Jumbo dynamos, generators Edison had named after P. T. Barnum's gigantic elephant. On July 6, 1882, Edison threw the switch for a trial run of the first Jumbo. Three days later, with his employees assembled, he connected a second Jumbo dynamo to the first. This would be the final test of, and the ultimate reward for, all the years of labor. Edison described how the events unfolded as he brought his dream to life: "At first everything worked all right. . . . Then we started another engine and threw them in parallel. Of all the circuses since Adam was born, we had the worst then! One engine would stop and the other would run up to a thousand revolutions; and they would see-saw. . . . When the circus commenced the gang that was standing around ran out precipitately, and I guess some of them kept running for a block or two. It was a terrifying experience as I didn't know what was going to happen. The engines and dynamos made a horrible racket, and the place seemed to be filled with sparks and flames of all colors. It was as if the gates of the infernal regions had suddenly opened." Edison and E. H. Johnson, the only two men in attendance who retained their presence of mind, turned off the engines and began the painstaking process of identifying and repairing the problems.[39]

Finally, on September 4, 1882, Pearl Street was again ready for operation. The power station, in an anxious Edison's words, was "the biggest and most responsible thing I had ever undertaken. It was a gigantic problem, with many ramifications. There was no parallel in the world. . . . All our apparatus, devices and parts were home-devised and home-made. Our men were completely new, without central station experience. What might happen turning on a big current into the conductors under the streets of New York no one could say." After countless sleepless nights, the final product was at hand. Edison was nervous: "All I can remember of the events of that day is that I had been up most of the night rehearsing my men and going over every part of the system. . . . If I ever did any thinking in my life it was on that day."[40]

But the event proceeded without a hitch. John W. Lieb, the power station's chief electrician, threw the switch on the first of the six Jumbo generators. Edison, recognizing that the most important place for him to be was with the investors, personally turned on the 106 lamps of the offices of Drexel, Morgan, and Company at 23 Wall Street.[41] The *New York Times*, whose offices also entered the electrical age that day, observed, "It was not until about 7 o'clock, when it began to be dark, that the electric light really made itself known and showed how bright and steady it was." A *Times* reporter described it as "soft, mellow, grateful to the eye; it seemed almost like writing by daylight." On opening day, Pearl Street served eighty-five customers with a total load of four hundred incandescent lamps.[42]

As Pearl Street grew, so did Edison's ambition. When it first opened, Pearl Street was capable of powering 1,616 lamps. By the end of its first year of operation, capacity had grown to 11,555 lamps.[43] Although the power plant lost money the first five years, Edison had proven that it was possible to operate a central station on a commercial basis.[44] If Edison wanted to expand, however, he would have to manufacture the components that his system required. Unfortunately, the stockholders of the Edison Electric Light Company—he no longer owned the majority of shares—wanted to limit its business to the licensing of power plants.[45]

Edison's willingness to invest his own money in manufacturing distinguished him from other investors and was largely responsible for his reputation as a great entrepreneur. To finance his foray into manufacturing, Edison sold much of his stock in another enterprise, the Illuminating Company. The Edison Lamp Company was established at Menlo Park to build incandescent lamps. In New York City, the Edison Machine Works was founded to build generators for the central stations, and the Edison Electrical Tube Company to make underground conductors. Edison also invested in Bergmann & Company, a small manufacturer of electrical fixtures.[46]

During the 1880s, the Edison system of electric lighting and power distribution spread throughout America and the world.[47] In addition to four installations in New York City, large power stations were located in Detroit, New Orleans, St. Paul, Chicago, Philadelphia, and Brooklyn. Overseas, London, Berlin, and Milan all had large central stations. Privately owned Edison plants, like those installed in the Vanderbilt and Morgan households, were supplying electricity to individual customers in Manchester, Paris, Amsterdam, Munich, and Rome, among others.[48]

Edison's fame spread even faster than his system of electric lighting. After a visit to his New Jersey laboratory, a reporter dubbed Edison the Wizard of Menlo Park. The name stuck, for it seemed to capture the magic and sense of wonderment that accompanied the introduction of each new invention. Indeed, Edison is believed to be the inspiration for L. Frank Baum's *The Wizard of Oz*. The title character in the book was a self-promoting, well-intentioned midwestern wizard who was facile not only with audio and visual projection, but with electricity as well. Part of Edison's charm was that he was the first commercial inventor who promised to improve the living standards of the common man. Coming on the heels of major improvements in the telegraph and telephone, and, especially, the invention of a marvelous tinfoil talking machine called the phonograph, the incandescent lamp forever established Thomas Edison's reputation as a national hero.[49]

By the end of the 1880s, Thomas Edison was on top of the world; he was the preeminent American inventor. Myths sprouted up around him. He was portrayed as a genius who toiled away in his laboratory for long hours as the world slept. While Edison did work extremely long hours, accounts of his several-day-long vigils were often greatly exaggerated. Described by the popular press as possessing "extraordinary propensities for work" and an "indomitable perseverance,"[50] Edison promised the

(© Bettman/Corbis)

George Westinghouse

world a new patent every ten days and a major invention every six months. Often described as a "giver of light," Thomas Edison did indeed serve to enlighten the world, both with his revolutionary incandescent lamp and his other marvelous inventions.[51]

EDISON'S DOMINANCE of the electrical industry did not go unchallenged. George Westinghouse, the son of a machine-shop owner in Schenectady, New York, was destined to become Edison's chief rival. He was a stocky man; blunt, dynamic, a bit of a fop, with a walrus mustache and a taste for adventure. Westinghouse was a hard-driving businessman, but the antithesis of a robber baron. He did not believe it was right or even necessary to bribe politicians or cheat the public to be successful. In 1867, at the age of twenty-one, he invented a mechanical contraption for returning derailed railroad cars to the track and a reversible railroad frog—a switching device that allowed a train to cross from one track to another. These two products proved so durable that railroad companies rarely required replacements, and soon Westinghouse was searching for a new business enterprise.[52]

One afternoon while traveling from Schenectady to Troy, Westinghouse witnessed a train wreck. Although the day was clear and the roadbed was level and properly maintained, two freight trains crashed into each other.[53] The engineers saw each other, but the primitive, individually hand-operated mechanical braking system required about half a mile to stop. After examining the wreckage, Westinghouse concluded that there was money to be made preventing railroad accidents. A year later, he invented the air brake, an automatic braking system powered by compressed air.[54]

Westinghouse spent much of the 1870s in England peddling his new air brake and building factories to serve the European market. While overseas, he became interested in railroad signaling and switching devices. He also gained valuable experience in manufacturing and marketing—knowledge that would be of great assistance with his eventual investment in electrical engineering. When Westinghouse returned to America in 1881, he established the Union Switch and Signal Company in Pittsburgh. Within a year, he began manufacturing and marketing automatic signal and switching mechanisms.[55]

Pittsburgh was rich in natural gas. Before long, Westinghouse drilled a well in his backyard and began selling gas to residential and commercial customers. In Pittsburgh, he pioneered the transfer of natural gas over long distances. By widening the pipe near the place of gas consumption, Westinghouse was able to use the high pressure at the well to convey gas under reduced pressure to customers several miles away. This simple method of distribution, later applied to electricity in the form of step-down transformers, was to account for Westinghouse's eventual dominance in the electrical lighting industry.[56]

Unlike Edison, Westinghouse did not get into the electrical business by manufacturing and marketing his own inventions. Instead, he bought up available patents and hired a skilled staff of engineers to work on making practical improvements.[57] Westinghouse first entered the field of electricity in 1884 when he purchased a direct current, self-regulating dynamo from William Stanley, an electrical engineer whom Westinghouse's younger brother, Herman H. Westinghouse, met on a train. The Stanley generator was a major advance. The dynamos produced by the Edison Company had to be regulated by hand in order to keep the current proportioned to the demand placed upon it. Otherwise, turning off a light would cause those remaining on to receive additional power, thereby jeopardizing the functioning of electrical appliances.[58] The Stanley generator was automatic.

Still, there was little economic advantage in the new Stanley generator, and in the beginning Westinghouse was not a serious competitor. His main concern remained the expansion of his gas business. Then, in 1885, Westinghouse read an article in Engineering, the leading London technical journal, that detailed an alternating-current system on exhibit at the Inventions Exhibition in South Kensington. The new system featured a Gaulard-Gibbs secondary generator, or induction coil, which was designed to transform high voltage into low voltage without loss of current.[59]

Westinghouse immediately saw the possibilities. Both the Westinghouse and Edison systems of direct current had one major drawback: the economical long-distance transmission of electrical power necessitated high voltage. But high voltage was too powerful and dangerous to be of use in the home or office for incandescent lighting. It had to be divided and stepped down at the place of consumption. The Gaulard-Gibbs induction coil or transformer did this, but it only worked with alternating current. It was not until well into the twentieth century that a transformer was developed for direct current.[60]

For $50,000, Westinghouse purchased an option on the U.S. patent rights to the Gaulard-Gibbs transformer.[61] He then instructed William Stanley to determine whether it was suitable for use in a central power station. After a brief examination, Stanley reported that it had little commercial value. The method of connecting the transformers was not practical, resulting in poor voltage regulation and the risk of complete system failure.[62] A new transformer would have to be engineered.

Stanley began this task in Great Barrington, Massachusetts, a small town in the Berkshires where he had vacationed as a boy. Under a restrictive contract agreement, Westinghouse furnished Stanley with a complete laboratory, paid him a generous salary, and covered all expenses up to $200 a month. He also gave Stanley one-tenth of the stock in the soon-to-be-formed Westinghouse Electric Company.[63] In return, Stanley agreed to surrender his rights to all commercial products he might invent. Stanley could only market the inventions to other businessmen if Westinghouse decided not to patent his inventions.[64]

In the autumn of 1885, Stanley leased an abandoned rubber mill at the north end of Great Barrington. While he waited for a Siemens alternator to arrive from London, Stanley began construction on twenty-six trans-formers.[65] The transformers were to be placed in the basements of the buildings to be lighted, and kept under lock and key so that competitors would not be able to examine them.[66] Stanley also hired a local man to solicit customers for his new alternating-current lighting system.[67]

On March 20, 1886, Stanley held a public demonstration. Copper wires were strung from the old rubber mill to the center of town, a little less than a mile. The conducting wires were attached to insulators nailed to the grand old elm trees that lined the main street of Great Barrington. In the early evening, Stanley illuminated thirteen stores (including his cousin's grocery), two hotels, two doctor's offices, one barbershop, and the telephone and post offices.[68] The era of alternating current had begun. Electrical history was transformed.

After further testing in and around Pittsburgh, Westinghouse opened the first commercial central station manufacturing and distributing alternating current in Buffalo, New York, on the day before Thanksgiving, 1888.[69] Electrical current was transmitted at 1,000 volts and delivered to the consumer at 50 volts, a ratio of 20 to 1. This made it possible to transmit electricity economically over great distances and still deliver current at a safe and useful voltage. Unlike direct current, customers were no longer required to live within a few hundred yards of the central station to obtain electrical service.[70]

The Westinghouse system was an immediate success. Within a few months, twenty-five orders for new central stations had been taken. By the end of two years, Westinghouse had installed his system in 130 cities and towns.[71] Edison sales agents became alarmed, but Edison himself remained confident that direct current was superior to alternating current. Edison had just introduced a three-wire system using a feeder main that increased almost threefold the voltage at which direct current could be transmitted.[72] And he was hard at work on a direct-current transformer.[73]

Despite its advantages, the Westinghouse AC system had two major drawbacks. It had no meter to measure current, nor did it have a motor to power machinery. Without a meter, Westinghouse was forced to charge each customer a flat rate based on the number of incandescent lamps. Since customers' bills did not depend on the amount of electricity consumed, they tended to let lamps burn continuously, rarely turning them off. This greatly increased the amount of electricity required from a central power station. When meters were introduced in August 1888, electrical consumption dropped by almost 50 percent.[74] This meant that output could be cut in half while earnings remained stable. It also meant that the same central station could serve twice as many customers and make twice as much money.[75]

Similarly, the lack of an AC motor to power machinery impeded the growth of the Westinghouse system. This problem proved more difficult to solve. Still, six months before the Buffalo central station opened for business, Nikola Tesla patented a primitive AC motor. Although it was not until 1892 that a practical AC motor was ready for sale, word that Westinghouse had engaged the exceptionally talented Tesla to improve his motor helped convince potential customers that a workable motor would soon be available to them.[76]

For the next few years, Westinghouse continued to gain ground on Edison. By the spring of 1890, the Westinghouse Electric and Manufacturing Company had sales of more than $4 million.[77] New clients were being taken on every day, and Westinghouse was well on his way to replacing the firmly entrenched DC with its more efficient cousin, AC. Westinghouse's path to eventual dominance of the electrical power industry would not be smooth, however, because Thomas Edison had more than money at stake, and his sense of pride and probity ran deep.

AT A VERY EARLY AGE, Thomas Alva Edison showed an eye for opportunity. He was a genius at recognizing consumer demand—and he knew how to exploit it. While still a boy, Edison published his own small newspaper, the *Weekly Herald*, for passengers of the Grand Trunk Railroad. He became fascinated with the power of the media to inform and influence; it was an experience that he would later cultivate in media blitzes for his numerous inventions. Stories tell of "Little Al's" (as he was called) childish fascination with electrocuting insects and rodents, and his irrepressible curiosity that, from time to time, landed him in trouble. One anecdote tells of a young Edison setting fire to his father's barn "just to see what it would do," as he later explained.[78] It was this mischievous curiosity coupled with his relentless pursuit of knowledge that propelled Edison toward greatness.

The Wizard of Menlo Park did more than invent, develop, and manufacture. He captured a nation's imagination, and to this day remains an American icon and legend. Nonetheless, Edison was subject to the same foibles and imperfections as any person. He was often known to remark with a false modesty that "genius is one percent inspiration and ninety-nine percent perspiration."[79] While there can be no doubt that Edison was an inventor without equal, he rarely credited the researchers who did much of his work and without whom he would have accomplished much less.

Though he was no anomaly in the unruly industrial age, Edison certainly relied on more than his ingenuity and scientific skill to launch his distinguished career. He was known to remark about his competitors, "I know how to steal. They don't know how to steal." In a cutthroat industry, Edison was as competitive as he was inventive. As he once put it, "I don't care much for the fortune . . . as I do for getting ahead of the other fellow."[80]

In accordance with his modest position in American industrial history, George Westinghouse played an indirect role in the battle of the currents. He was born in October 1846, making him just four months older than his more famous adversary, Thomas Edison. Westinghouse's appearance, in sharp contrast with Edison's, was generally described by his peers as "distinguished." He stood well over six feet tall, had a strong build, and sported a walrus mustache. Like Edison, Westinghouse was a tireless worker, a man of modest appetites and measured habits. Westinghouse and his wife took pleasure in entertaining; they frequently hosted dinner parties and dined with such eminent luminaries as Lord Kelvin, Earl Grey, and at least one American president. A veteran of the Civil War, Westinghouse had only three months of college, and, as his official biographer has written, "He did things that the text books said were against the laws of nature, and, in course of time, the text writers caught up with him." Of course, he also did things contrary to the advice of the textbooks "that his engineers protested against, that cost time and money, and that ended up on the scrap heap."[81] Westinghouse was, by all accounts, a man of intelligence and integrity. Though he did not have Edison's ability to attract attention to himself, he was a shrewd businessman and a capable engineer.

The difference between Edison and Westinghouse is obvious when seen against the backdrop of their respective encounters with Nikola Tesla. Tesla, who worked for an Edison subsidiary in Paris, had an enthusiastic letter of introduction to Edison from Charles Batchelor, as well as a working knowledge of engineering and electricity. Edison hired Tesla, who promptly proposed an elaborate plan for redesigning Edison dynamos to make them more efficient and less costly. The project would take a long time, but Edison, tempted by the young Serb's promise, told Tesla that there would be a $50,000 bonus for him if he succeeded. Tesla toiled feverishly for the better part of a year until he finally presented Edison with a successfully redesigned dynamo. However, when he asked for his $50,000, Edison replied "Tesla, you don't understand our American humor." After Edison offered him a $10 raise on his $18 per week salary, Tesla, angry and disgusted, put on his bowler hat and walked out.[82]

After leaving Edison's laboratories, Tesla patented an AC motor and soon found himself working for George Westinghouse. The two men instantly developed a strong rapport, and Westinghouse, recognizing Tesla's extraordinary talent, offered him a generous contract complete with royalties. But as the Westinghouse Company began to feel the financial burden of competition with J. P. Morgan, who was Edison's primary backer, and as Tesla's patents came into greater and greater use, investment bankers warned Westinghouse that his contract with Tesla needed to be terminated if he was to remain fiscally solvent.[83]

Westinghouse resisted. He believed in royalties, but his business sense told him that the contract must be broken. Westinghouse solicited Tesla's compliance, acknowledging that the contract had undoubtedly been made in good faith. Yet, he implored Tesla to let him break it, arguing that the entire Westinghouse Company depended on it. Tesla acquiesced, saying, "Mr. Westinghouse, you have been my friend, you believed in me when others had no faith; you were brave enough to go ahead . . . when others lacked courage; you supported me even when your own engineers lacked vision to see the big things ahead that you and I saw; you have stood by me as a friend." The Westinghouse Company paid Tesla $216,600 for the outright purchase of his patents, and proceeded to make Tesla's 60-hertz AC system the standard for the electrical power industry.[84]

Initially, Edison did not consider Westinghouse a threat. He wrote to one of his many associates: "None of Westinghouse's plans worry me in the least; the only thing that disturbs me is that Westinghouse is a great man for flooding the country with agents and travelers. He is ubiquitous and will form numerous companies before we know anything about it."[85] But as Westinghouse continued to champion the cause of alternating current in the mid-1880s, Edison became increasingly concerned. He launched a patent war against all electric companies, with Westinghouse as his primary target.[86]

On December 23, 1887, Edison General Light Company initiated eleven lawsuits against the Westinghouse Company. An Edison associate told *The Electrical World*, "Mr. Edison has chosen to bring these infringements of his patents against the Westinghouse people because they are believed to have done more toward the infringement of his patents than any of the other companies. This litigation will certainly result in a complete victory for Mr. Edison and this country."[87] For his part, Westinghouse told the Pittsburgh Dispatch, "[The suits were] news to me. However I am not surprised at the action of the Edison people. It is all bunkum on their part. We have patents on our systems of incandescent lighting that are so immeasurably superior to theirs it is only a question of time before we supersede them all over the country. Our systems are entirely different from theirs, and in no way are they an infringement on the Edison system."[88]

One of the most significant suits involved the hermetically sealed chamber that creates the vacuum for the incandescent lightbulb. Edison and Westinghouse attorneys argued in the U.S. Circuit Court of Appeals for weeks, and Edison's lawyer embellished his arguments with very grave, sweeping rhetoric. "These people," he said, "have absolutely nothing to stand on. They have appropriated every idea of Mr. Edison's and all this without the slightest compensation. They say in defense that Edison did not make the discovery, or that he did not perfect it, or that it was neglected, or that he delayed too long in bringing the original suit. What rot!" Of Westinghouse, Edison said, "The man has gone crazy and is flying a kite that will land him in the mud sooner or later."[89]

Edison also tried to attack Westinghouse on a legislative level. He lobbied Albany senators to pass a law limiting electrical currents to 300 volts.[90] This would effectively eliminate AC's advantage, because it had to run at a much higher voltage to be of any practical use. Edison's arguments convinced at least one powerful politician: former New York governor Alonzo B. Cornell wrote several letters to the then-current city mayor, A. S. Hewitt, attempting to convince Hewitt to outlaw high-tension circuits, thus negating AC's competitive edge. Westinghouse, however, threatened to sue the Edison firm and others for conspiracy under the laws of the state of New York, and so the legislators were not persuaded.[91]

Edison tried again. In 1887, a similar bill was introduced in the Virginia State Senate. A committee of fifteen senators was appointed to hold public hearings on the matter. Edison himself testified in the hearings and was the first witness called in support of the bill. He was questioned by lawyers and members of the committee, but Edison, who was hard of hearing, had trouble deciphering the queries, which frequently had to be repeated. It was one of the few times that Edison publicly failed to carry the day. An engineer present at the hearings recalled Edison's testimony as "neither constructive nor convincing," although it was made clear that he favored the bill.[92] The bill failed.

As the private animosity between Westinghouse and Edison became more public, other people began to take advantage of the situation. The spoils of the Westinghouse and Edison war went chiefly to those who sought to capitalize on them both. Edison's chief associate, Charles Batchelor, received this letter from businessman P. B. Shaw of Williamsport, Pennsylvania, dated September 27, 1889:

Dear Batchelor:

I enclose you herewith copy of a patent for an Electric Car Brake. Mr. Hinkley the inventor is a resident of this city, it has occurred to him that in view of the competition between the Edison Electric light Co., and the Westinghouse Elec. light Co., that possibly the Edison people might feel disposed to look into the matter, that he Hinkley thinks will knock air brakes sky high. I have no personal knowledge of the patent, never having seen the model, or read the specifications, but I know Mr. Hinkley and if he is as level on this subject as he is in his usual business dealings he must have a good thing. I send this to you, knowing that if it has any merit you will see that the subject is properly investigated. Will you please write me what you think of it, and what consideration if any you will give to him.

Very truly yours.

P. B. Shaw

P.S. The Westinghouse people are building a plant in this city, which they claim will have twice the capacity of our Edison Co. and double the economy: I shall soon be looking for a job.

P. B. S.[93]

It was not long before the Westinghouse Company took the lead in the race to electrify America. Longtime Edison customers wrote him reluctant apologies and then switched to the Westinghouse system. Edison's own salespeople began to appeal to him to do something. One of Edison's agents, W. S. Andrews, wrote to Edison on May 21, 1887, with a message of desperation, telling Edison of how Westinghouse was supplying electricity at a much cheaper cost and allowing unlimited hours of burning. Andrews reported that the agents were feeling very "blue" about it, and that though "no company will be so foolish as to continue such a suicidal policy of selling at a loss," he still feared for the local company's survival. "It is very much to be regretted," Andrews wrote, "that the Edison Company have not taken more active steps to provide local companies with some means of combating with such as Westinghouse in the matter of long distance lighting." He urged Edison to "give [us] a few crumbs of comfort" in these desperate times.[94]

Another Edison agent, W. J. Jenks, wrote to Edison and voiced a common concern among Edison employees: "Are we going to sit still and be called 'old fashioned fossils' &c., and let the other fellows get a lot of the very best paying business?"[95] On October 30, 1889, sales agent J. H. Herrick delivered the ultimate indignity. He suggested that Edison develop "an alternating system which will follow as closely as possible the model of the Westinghouse system."[96]

Throughout all of the turmoil, Westinghouse and Edison had very little contact with each other. On June 7, 1888, however, Westinghouse wrote a conciliatory letter to Edison attempting to rebuild their relationship. He even suggested that the two companies bury the hatchet and work together. "I believe there has been a systematic attempt on the part of some people to do a great deal of mischief and create as great a difference as possible between the Edison Company and the Westinghouse Electric Co., when there ought to be an entirely different condition of affairs. I have a lively recollection of the pains that you took to show me through your works at Menlo Park when I was in pursuit of a plant for my house, and before you were ready for business, and also of my meeting you once afterwards at Bergman's factory; and it would be a pleasure to me if you should find it convenient to make me a visit here in Pittsburgh when I will be glad to reciprocate the attention shown to me by you." In this same letter, Westinghouse declined to join a trust proposed to him by Dr. Otto A. Moses, but said that he would consider a Westinghouse/Edison merger.[97]

However, on July 3, 1888, Westinghouse sent Edison a letter of a much different tone, and enclosed in it a confidential memorandum he had acquired that Edison had sent out to his company. The memorandum stated, "If the Westinghouse Company, as they claim, have a system so superior and so much cheaper than the Edison, and if, as they pretend, the Edison Company's patents are valueless, why do they desire to bring about a combination with the Edison Company?"[98] The memo, which Westinghouse had gotten his hands on, had angered him. He, like Edison, was now more interested in making war than peace.

Edison's hostility toward Westinghouse was obvious. Also obvious was the fact that the Edison Company was growing more and more desperate to halt the expansion of alternating current. Although Edison never went through with his plan, he had considered developing and marketing his own AC system. He never intended it for sale, though, because after he had extensively advertised his new system, he was going to withdraw it—telling the public it was unfit for use and that he could not in good conscience sell it. Since it would have been identical to the Westinghouse system, the damage would surely have been great, but Edison, perhaps not willing to stoop so low, never carried out the scheme.[99]

Unable to hinder AC's progress through legal channels, Edison finally used his most valuable asset: the public's trust. When Thomas A. Edison, the Wizard of Menlo Park, spoke, people listened. The most effective phase of Edison's attack on Westinghouse came when Edison began to speak out about the "dangers" that he felt were inherent in alternating current. He began slowly, first planting the seeds of doubt in the hands of top men in the electrical power industry. Many years earlier, in anticipation of Westinghouse's potential success, he had written to E. H. Johnson, "Just as certain as death Westinghouse will kill a customer within six months after he puts in a system of any size. He has got a new thing and it will require a great deal of experimenting to get it to work practically. It will never be free from danger."[100] As Edison became more intent on destroying Westinghouse, his tactics became more desperate. One of his employees at his West Orange laboratory reported, "As Edison saw it, accidents caused by AC must, if they could not be found, be manufactured, and the public alerted to its hazards."[101]

And alert the public he did. In February 1888, Edison published a bright red pamphlet entitled "A Warning from the Edison Electric Company."[102] Outwardly his aim was to correct, for the sake of the general public, the many misstatements and misrepresentations of the George Westinghouse Consolidated Electric Light Company and the Thomson-Houston Electric Light Company, another AC competitor.

Edison began his pamphlet by reminding the reader that this was not their first warning. Indeed, he had issued an earlier caveat in the form of a card that was distributed in May 1885, which, in a thinly masked threat, reminded readers that Mr. Thomas A. Edison held the exclusive patent for the manufacture, sale, and use of any and all practical incandescent lamps. Even more bluntly, the card asserted that the Edison Company would continue its "vigorous prosecution" of all patent violators, including George Westinghouse and Thomson-Houston, whom Edison regarded as having stolen inventions that rightfully belonged to him.[103] Edison's card stated that his was the only company that could guarantee the protection of its customers from lawsuits, whereas his competitors were placing their customers in legal and financial jeopardy because of their unauthorized use of Edison's incandescent lamp.[104]

Finally, Edison warned readers that there was great danger in alternating current that was being ignored by "advocates of cheapness." He pointed to the "glorious record" of his own system, which had never caused a fatality, and speculated that the questionable safety of AC would end in its being "legislated out of existence."[105]

Edison kept up the attack. On October 16, 1889, the *New York Evening Sun* declared in a headline, "Edison Predicted It." "It" was the New York Board of Health's conclusion that the only way to light safely by electricity was to control, by ordinance, the strength of the current. The paper quoted an Edison speech from 1884 when he declared, "Unless the electric light companies reorganize their systems, so as to reduce the tension on conductors, there is going to be trouble. The tension will become dangerous. It will kill men, and something like panic will follow. But to reduce their tension it will be necessary for companies to increase the amount of copper used in the construction of their dynamos and generators to enlarge the conductors on their circuits. But it is not likely that they will do this as it will be much more costly than the present system. Some day they will have to come to it, for the high tension current is fearfully dangerous, and will begin to kill." The paper concluded, "So here we have the hard fact about it. The companies used cheaper dynamos and cheaper construction, thereby generating a deadly current, when at a greater cost safety could have been secured. It is the fault of the system, not the wires. Mr. Edison sees, as the community now sees, that safety can alone be guaranteed by ordinance, regulation and inspection."[106]

Edison's single-minded determination to hammer home the deadly nature of AC began to echo in all his public statements. His conviction is expressed explicitly in reply to a letter that inquired about how much voltage the human body could endure. Any voltage could be applied, Edison responded, because "it's the current that kills,"[107] a true, but nonetheless misleading, answer.

Neither Westinghouse nor Edison was a stranger to the power of the press. Edison's early training with his own newspaper had taught him the value of print media. Both the Edison and Westinghouse camps flooded newspapers with editorials and "informative" articles about their respectively preferred currents. In an article in the November 1889 issue of *Public Opinion*, Edison sent this message to the public:

> The currents used for electrical lighting at the present time may generally be divided into four classes: First, the low-tension continuous current, with pressure not exceeding 200 volts, used for incandescent lighting; second, the high-tension continuous current, with a pressure of 3,000 volts and over; third, the high-tension semi-continuous current, with a pressure of 2,000 volts and over; fourth, the alternating current, with a pressure from 1,000 to 3,000 volts and over. The first is harmless, and can be passed through the human body without producing uncomfortable sensations. The second is dangerous to life. Momentary contact with the conductor of the third results in paralysis or death, as has frequently occurred; and the passage of the fourth, or alternating current through any living body means instantaneous death. These are the simple facts which can not be disproved.[108]

BEFORE THOMAS EDISON's fierce, often irrational opposition to alternating current can be appreciated, it is necessary to understand the basic difference between direct and alternating current. Direct current can be thought of as water in a pipe flowing in one direction. This movement of electrons produces an electrical current. Alternating current can be thought of as water in a pipe flowing in one direction, then reversing itself and flowing in the opposite direction. Under the U.S. standard, AC does this sixty times a second. The current consists of the back-and-forth motion of the electrons. In direct current, there is a net, very small motion of electrons in one direction. In alternating current, the electrons remain in place. AC and DC produce identical effects, since it is the movement of electrons, not their direction, that produces the electrical current.

If AC and DC are essentially identical in their ability to perform work—the differences lie in the manufacture, delivery, and utilization of current—why did the pragmatic Edison resist so adamantly the advent of alternating current?

Despite his obvious proprietary interest in DC, Edison could have easily changed over to AC. In November 1886, at the urging of European agents, the Edison Company purchased the American rights and options to the Hungarian-manufactured Zipernowsky-Blathy-Deri (ZBD) alternating-current system. "These patent rights and options gave the Edison Company the opportunity to take up alternating current with patent protection so strong that Westinghouse, depending on the Gaulard and Gibbs series patent, might have been forced out [of business] completely."[109] To this end, and at the urging of the Edison Company president, E. H. Johnson, the eminent electrical engineer Frank Sprague prepared a confidential, in-house report on the feasibility of changing over to alternating current. Sprague recommended that Edison switch to alternating current. It was significantly more economical to transmit over long distances, and the future of the electrical power industry lay in long-distance transmission.[110]

Even if Edison could not bring himself to follow completely Sprague's recommendation, he could still have pursued a method of utilizing alternating current for long-distance transmission. All he really needed to do was develop a device that converted direct current into alternating current for transmission and then reconvert it back to direct current for local distribution. This would have allowed Edison to retain his reliable and highly efficient direct-current generators, as well as to protect the investments of licensee companies in local distribution.[111] Although this strategy might have allowed Edison to retain his dominance over the electrical-power industry, it was Westinghouse who first developed the rotary converter in the early 1890s. There is no evidence that the Edison Company ever attempted to design or build an AC/DC converter.

So why didn't the Edison Company gradually convert to alternating current? In industrial history, it is an old question, one that has been asked many times. In some sense, the answer is no more complicated than that Edison was in the direct-current business. Corporations in an established industry often fail to recognize the value of a new, competing product until it is too late. The gas companies, for example, did not recognize the potential value of electricity, the railroads were not concerned about the automobile, and typewriter and adding-machine companies failed to anticipate the ascent of the personal computer. A few critics have suggested that Edison lacked the intellectual ability to understand alternating current. He had little formal schooling, none in science, mathematics, or electricity, and he could not comprehend the scientific principles involved in designing alternating-current systems. Even in electricity and chemistry—two areas he knew best—theory and abstraction were beyond his comprehension. Once, he all but admitted that he did not understand alternating current: "How can they reverse the direction of the current?"[112] Edison is said to have asked. Many years after having left Edison's employ as a young laboratory assistant, Francis Jehl described Edison as a "man that cannot solve a simple equation."[113] Recently, much has been made of Edison's alleged dyslexia. He has become a shining example of an individual with learning disabilities who still managed to accomplish great things. However, a review of Edison's personal papers and laboratory notebooks tends to discredit this opinion.[114]

For the man who invented the phonograph, the incandescent lightbulb, motion pictures, and the alkaline storage battery, and for the man who holds 1,093 patents, more than anyone else in history, an explanation that relies on his intellectual limitations hardly seems satisfactory. Instead, we might look to the makeup of his character and personality to explain his inability to recognize and embrace alternating current.

By all accounts, Edison was a stubborn man. He had a strong belief in himself, always thinking he could succeed where others had failed. Often iconoclastic and usually audacious, Edison tended to dismiss the opinions of others, especially when they clashed with his own. For example, Nikola Tesla tried in vain to persuade Edison that the future belonged to alternating current. Although he acknowledged Tesla's unusual talent, Edison often ridiculed him by calling his ideas "magnificent but utterly impractical."[115] Edison's conceit even extended into areas he did not entirely comprehend. In a patent suit brought by Westinghouse, Edison proudly admitted in open court that he did not fully understand Ohm's law. "Such knowledge," he added, "would prevent me from experimenting."[116]

Edison's unshakable belief in the superiority of applied science induced him to disdain knowledge for knowledge's sake. He felt that a scientist who spent his days on theoretical problems without an anxious eye toward their practical application was wasting his time and talent. He liked to boast that he was a successful inventor because he was not corrupted by schooling. For Edison, practical application and marketplace success were the main measures of achievement. He often expressed the view that mathematics was of little value in guiding research. "The mathematics always seems to come after the experiments—not before."[117] He was fond of announcing, "I can hire mathematicians, but they can't hire me."[118]

Most people reach their psychological limitations long before they reach their intellectual ones. Thomas Alva Edison was no exception. His hatred for George Westinghouse, his ego investment in the industry he created, and old-fashioned pride all made it difficult, perhaps impossible, for Edison to act as the shrewd entrepreneur he was. His cutthroat competition with Westinghouse would test the limits of Edison's character, leading him to risk his reputation as a national hero and ultimately to betray the public's trust. In some sense, the answer to why Edison, the great innovator and visionary, did not embrace AC is no more complicated than the fact that it was not his—he did not invent it.

THE STRUGGLE between Edison and Westinghouse cannot, by itself, explain how the electric chair came to replace hanging as the preferred method of execution. As popular and influential as he was, Edison did not create the electric chair out of thin air; instead, he tapped into an ongoing social and political movement and shaped it to suit his own pecuniary needs. The American ideals of social progress and scientific advancement gave rise to a growing uneasiness with public displays of barbarity. During the last quarter of the nineteenth century, New York came to realize that it needed to modernize its method of execution if capital punishment was to survive into the next century. Edison needed only to lend his prestige as an inventor without equal, and his celebrity, to the effort to bring about a much-needed change in the administration of capital punishment.

CHAPTER THREE

The New Electrical Execution Law

IN THE EARLY PART of the nineteenth century, executions in New York, as elsewhere in America, took place in public.[1] They were elaborate spectacles designed to act as object lessons in deterrence for those who might otherwise contemplate a life of crime. People took their children.[2] Unlike modern executions, the condemned prisoner played a central and meaningful role. Amid much pomp and circumstance, the condemned was paraded from the jail to the gallows. As the wooden ox-driven cart carrying the prisoner, followed by a long procession of spectators, creaked and rolled slowly toward the hanging place, solemn music was played. Once there, the condemned was greeted by friends, public officials, spiritual advisers, and a crowd of curious onlookers. Religious leaders delivered sermons on the causes of crime and its deadly consequences, the terrible "wages of sin." The condemned frequently repented, asking the forgiveness of his neighbors, warning them not to follow in his fatal footsteps. Although there were always a few dissenters, the execution served to unite spectators, local officials, and the condemned in an affirmation of the laws, values, and moral principles of the community. It brought people closer together.[3]

THEN THINGS began to change. Prior to the 1830s the condemned was usually a stranger, an outsider with few if any ties to the community. As the population soared and urbanization took effect, however, the condemned was now likely to be a member of the community, and the officials themselves representative of outside authority. Because the condemned was regarded by many as more of a martyr than a criminal, the harsh punishment tended to weaken social cohesion. The doomed men and women often turned their singular moment of shame and piety into an opportunity to denounce the officials responsible for their conviction and impending death. Religious leaders also soured on public executions. They regarded last-minute gallows conversions as disingenuous, a sham that mocked rather than honored the sanctity of the hour. And, finally, the large crowds, on occasion numbering more than thirty thousand people, regularly became lawless.[4] The violent passions of the so-called dangerous classes, regularly excited by the spectacle of hanging, led increasingly to rampant drunkenness, ribaldry, and, on occasion, attempts to rescue the condemned.[5]

It was during this time that the public acquired an awareness that punishment, especially grave public punishments like hanging, did not work; that they may even be counterproductive to the preservation of social order. Penal reformers spoke out against the brutal and bloodthirsty wave of crime and violence that permeated the community on execution day. As crime flourished, undeterred by the "lessons" of public hangings, reformers searched for more humane and less barbarous methods of punishment. About the same time, the moral ground shifted, as evidenced by the rise of philanthropic societies dedicated to social and humanitarian causes.[6] And, for the first time, rehabilitation became an expressed goal of punishment. It was during this era that the great custodial institutions— penitentiaries, insane asylums, and almshouses—were built.[7]

Frequently bungled executions also led to a widespread distaste for public hanging. It was the county sheriffs' responsibility to execute criminals sentenced to death within their jurisdiction. Since few possessed the concrete knowledge and experience of a skilled hangman, botched executions were common.[8] Under the technique employed in New York, the condemned was led to the gallows, where the knot of the noose was placed just under his left ear. The other end of the rope was secured to a metal weight approximately twice the mass of the condemned criminal's body, hanging four or five feet in the air by a guard rope. When signaled, the sheriff would cut the rope, causing the metal weight to fall and the body to be suddenly and violently yanked into the air. If everything went as expected, the condemned's neck was snapped and death followed instantaneously. If the knot was not properly adjusted or if the man turned his head, then a slow, agonizing strangulation occurred as the condemned dangled feebly for twenty to thirty minutes. Conversely, if too much weight was placed on the counterbalance, or if the relationship among the length of fall, the weight of counterbalance, and the weight of the condemned's body was miscalculated, the unexpectedly brutal snap of the rope would tear the head off the body.[9]

With each botched hanging came a renewed call for the abolition of capital punishment. As early as 1832, the New York Assembly appointed a committee to "inquire into the expediency of a total abolition of capital punishment." Silas M. Stilwell of New York City headed the committee, which argued that it was not necessary to take the life of a murderer and that keeping him in confinement to prevent future crimes was more than adequate. The committee believed public executions were not a significant deterrent to crime and that the awful displays of cruelty and suffering at the gallows tended to promote, rather than deter, crime.[10] Stilwell's committee cited an incident in Lancaster, Pennsylvania, where a public execution ended with a dozen new serious crimes committed, some of which were capital offenses.[11]

Acknowledging that it was a powerful one, the committee attempted to explain away the biblical argument in support of the death penalty. According to the committee, this view had its roots in the text of the Old Testament, where God commanded Noah, "Whosoever sheds man's blood, by man shall his blood be shed." The committee dismissed this reasoning by virtue of its placement in the Old Testament, saying that the laws of the Jews "were made expressly for that people."[12] They argued that ancient Jewish law had no relevance to modern crime and punishment, and quoted several ministers who supported their position. Moreover, they argued that a sophisticated interpretation of the verse revealed further ambiguity. The clergymen concluded that the passage was more of a prediction than a law, in which case it should have no legal bearing whatsoever, even for ardent believers.[13]

Furthermore, the committee noted, the case of Cain and Abel dramatically illustrated God's approach to punishing those who have wronged their fellowman. "He certainly was a shedder of man's blood, and that in the most aggravated and depraved manner; but we do not find that the Almighty took blood for blood." Instead, they pointed out, God marked Cain to protect him from the wrath of the public. The committee wrote that God's protection of the murderer Cain from "the tendency of human passions to return injury for injury, blood for blood," clearly demonstrated that God did not favor these extreme methods of punishment, and, in fact, believed that murderers should be shielded from such violent retribution.[14]

Retribution—"an eye for an eye, a tooth for a tooth"—was also denounced by Stilwell's committee. They chronicled similar statutes in other governments, both ancient and modern, that addressed the punishment of vicious murderers but did not condone killing those murderers as penance or atonement for their crimes. They maintained the object of punishment was "not to gratify a spirit of revenge, for that would be perverting the current of justice." The commission's report then asked proponents of the death penalty to examine carefully their beliefs, claiming that any sentiments of revenge were unacceptable in the administration of the law.[15]

AFTER PROCLAIMING that vengeance had no place in the law, the committee addressed the problem of wrongful convictions. The report objected to the "irredeemable nature" of capital punishment. It noted that, despite all the safeguards written into the law, it was impossible to be "absolutely certain" that the condemned was indeed guilty. They further reported that there were more than one hundred capital cases in which the person executed was later proven innocent. The committee asserted that it was "better that nine guilty persons escape than that one innocent one should suffer." If life imprisonment were substituted for the penalty of death, Stilwell's committee argued, then the innocent person's case could be overturned and restitution made. Once the death penalty was administered, the wrongfully convicted remained beyond the reach of both justice and mercy.[16] An abolition bill was introduced but it went nowhere. Two years later, in 1834, Assemblyman Samuel Bowne, a Quaker lawyer from Otsego County, introduced another bill into the assembly to abolish capital punishment. It, too, was defeated, but this time by just four votes.[17]

ABOUT A MONTH before Bowne's bill was narrowly defeated, Erie County assemblyman Carlos Emmons introduced a bill to abolish public executions.[18] The Revised Statute of 1828 had given county sheriffs the authority to hang criminals within the confines of the prison or prison yard. The statute was designed to avoid the mayhem that often accompanied public executions. Because public hangings usually had the strong support of small-town shopkeepers and businessmen who regarded them as good for business, not a single sheriff chose to hold a private hanging.[19]

Emmons's bill was attacked from both sides of the aisle. Proponents of capital punishment wanted to retain public executions because they believed in their instructive and deterrent value. They felt that unless executions were held in public, some people might suspect that wealthy prisoners could buy their way out of a private execution. Those supporting the abolition of capital punishment thought executions should remain public, so that "their consequences and enormity might be more vividly impressed on the public mind." Conducting executions behind prison walls, they argued, would tend to remove the issue of capital punishment from the public consciousness, and thus be detrimental to the abolitionist movement.[20]

Although Emmons's bill died in the assembly, the following year the senate appointed another committee to study the question of public executions. Its leading member was Senator Ebenezer Mack. Within ten days, Mack's committee filed an eleven-page report. It began by noting that in earlier times, punishments were intended to be vindictive. The infliction of pain was thought to be essential to the goal of deterrence; terror was the agent of the law, and justice was unobstructed by mercy. Mack's committee rejected this argument as antiquated, advocating the more modern notion of reforming the criminal with the application of humane and rational principles of law and justice. These barbarous executions did not deter people from committing crimes, the committee argued, but rather they tended to "harden and brutalize" the public. Exposure to violent executions engendered more abhorrence than reverence for the law.[21]

In the last half century, Mack's committee pointed out, almost every nation of Europe had limited the number of crimes punishable by death and abolished primitive methods of inflicting capital punishment. Fewer death sentences, fewer executions, and, most important, a decrease in the level of crime were the results. The committee claimed that deterrence was not dependent on either the severity of punishment or its public nature, but on the moral education of its people. In the United States, whose laws are founded on the principles of humanity, the report noted, there was no room for public executions. The committee further found that any social function that capital punishment could provide had no lasting effect. It concluded that the death penalty was less effective in deterring criminal activity than the ever-present example of a man locked up in prison for the rest of his life.[22] The committee believed that a private execution would be much more "salutary and impressive" than a public one. In a public execution too much attention was focused on the condemned. During his confinement, he was besieged with friends and relatives, religious advisers, and newspaper reporters, all of whom hung on his every word. The committee maintained that the condemned often regarded all this attention as immensely preferable to dying at home alone in honest poverty. Instead of preparing himself for a public display of courage and bravado on execution day, the condemned man, faced with a private execution, would be more likely to spend his final hours reflecting on the "justice of his sentence" and to go to his death with a "broken and contrite heart."[23] Along with his report, Mack introduced a bill for the abolition of public executions. The senate quickly passed the measure; the assembly, which had defeated a similar bill two months earlier, passed this one by a margin of 66 to 20.[24]

AFTER THE ABOLITION of public executions in New York, the debate over capital punishment quieted down. Nonetheless, during the next thirty years, there were no fewer than six committee reports issued by the New York Assembly.[25] Although they varied in emphasis, each concluded that the death penalty had no place in a civilized society. Supported by the New York State Society for the Abolition of Capital Punishment, the bills that came out of these committees came close to changing the law but never succeeded in abolishing capital punishment. Public support for the reform effort reached its apex in 1846 and 1847. After that, New York City faced one crisis after another, as did the nation. Between 1825 and 1855 the city's population quadrupled. An influx of immigrants to New York City caused social and economic dislocation; the crime rate rose dramatically. City government was not equipped to handle massive social change. Blaming immigrants for the rise in crime, the public lost its desire to abolish capital punishment.[26]

As the Civil War approached, the effort to abolish capital punishment took a backseat to the moral issue of slavery and the political necessity of maintaining the Union. After experiencing the war's terrible carnage and loss of life, the American public developed a greater sensitivity to the pain and suffering of others. The hypodermic needle, first invented in 1832, was widely used on Civil War battlefields to inject wounded soldiers with morphine. At home, physicians began to administer drugs for the sole purpose of relieving pain; effective pain relievers began to flood the market, and people no longer regarded pain as an inevitable part of life.[27]

Press reports, which had previously limited themselves to euphemistic descriptions of the execution, began to describe hangings in vivid and often sensational terms. Reporters and columnists began to focus a critical eye on the process of execution, blaming officials for their cruelty and incompetence, and denouncing the crowds that still appeared outside the prison walls as being "morbid curiosity seekers."

AS DISSATISFACTION with hanging grew, a few scientists turned their attention to electricity as a possible means of execution.[28] On March 24, 1883, *Scientific American* published an editorial titled "Killing Cattle by Electricity." The journal argued that electricity might be the best way to kill "worn-out horses, asses, or even cattle used for food" and gave a detailed description of how this could be accomplished. The prestigious journal argued that its proposed technique would eliminate the "horrid circumstances now attending the slaughter of . . . animals" and provide a painless death.[29] It further maintained that until capital punishment was abolished, death by "judicial lightning" should be adopted in place of the "hideous violence" of hanging.[30]

Two years later, in another editorial, *Scientific American* returned to the issue of electrical execution. The journal approvingly quoted one of its readers, who asked, "What more scientific method can be devised than the application of electricity as an executioner?" The editorial proposed that criminals be executed by connecting the execution apparatus to nearby streetlights. "The criminal will be placed in a chair, with his head bound back against a bulb at the end of the wire, through which the fatal shock will be communicated," the journal explained. "And it would be possible to furnish the death seat with an automatic attachment so that the execution could be affected at a given moment by the action of a clock-like apparatus and without the least movement of the hand of the officer charged with the infliction of the death penalty." With this technologically advanced method, *Scientific American* asserted, death would be quick and painless, and the hangman could be replaced by the clock. This would eliminate the need for an executioner, and hence relieve anyone of direct responsibility for the death of the prisoner.[31]

The push to make executions more scientific and more humane led inextricably to the eradication of the rope. Dr. Alfred Porter Southwick, a Buffalo dentist, was the leading advocate for replacing hanging with electrical execution. As a young man, Southwick was an engineer for the Great Lakes Steamboat Company before becoming chief engineer for the Western Transit Company. In 1862, at the age of thirty-six, he left engineering to become a dentist. A founder of the New York State Dental Society, he later taught operative techniques at the University of Buffalo Dental School.[32] Southwick, a Quaker by upbringing, was a stern-looking man who wore an expression of prophetic sapience. He was active in the eugenics movement and, like many of his peers, understood himself to be a visionary for an improved human race.[33]

Southwick's interest in electrocution can be traced back to an incident in 1881, when he witnessed the accidental electrocution of a Buffalo drunkard who had stumbled onto a live wire.[34] Since the man appeared not to suffer, it occurred to Southwick that electricity might prove to be a quick and painless method of putting criminals to death. To test his idea,

(*New York Herald*, 7 August 1890)

Alfred P. Southwick

Southwick conducted a series of crude but deadly experiments on dogs in the Buffalo pound.[35] With great enthusiasm he concluded that electricity, this new, "mysterious fluid" that promised to improve the life of the average man, could also be a marvelously humane method of execution.

A few years earlier, Southwick had been instrumental in passing the first legislation regulating the practice of dentistry. And as a current member of the Board of Dental Examiners, he had influential friends in Albany, including New York state senator Daniel MacMillan. Shortly thereafter, Southwick began lobbying the New York Assembly in support of replacing the gallows with electrocution.

In 1884 New York governor Grover Cleveland had been elected president of the United States, and in January 1885 his lieutenant-governor, David Bennett Hill, succeeded him as governor.[36] With an election looming in the fall, Hill sought to capitalize on increasingly widespread anti-hanging sentiments. In his first message to the assembly, Governor Hill urged legislators to consider alternative methods of execution. "The present mode of executing criminals by hanging," Hill said, "has come down to us from the dark ages and it may well be questioned whether the science of the present day cannot provide a means for taking the life of such as are condemned to die in a less barbarous manner."[37]

Within a few days of Hill's address, Southwick's close friend, state senator Daniel H. MacMillan, successfully introduced a measure to form a committee to recommend a more humane method of capital punishment.[38] In response, Governor Hill appointed a commission, chaired by Elbridge T. Gerry. The grandson of one of the signers of the Declaration of Independence and the Articles of Confederation,[39] Gerry was a renowned New York attorney, counsel for the American Society for the Prevention of Cruelty to Animals, and a founder of the Society for the Prevention of Cruelty to Children.[40] Matthew Hale, a prominent Albany attorney and grandson of the Revolutionary War hero who uttered a famous epithet as he was about to be hanged by the British for treason: "I only regret that I have but one life to lose for my country," was also appointed a member of the commission. The third member was none other than the Buffalo dentist, Dr. Alfred Southwick.

The Gerry Commission met almost immediately, and began the long process of gathering information from experts in many fields, especially medicine and electricity. A survey was sent to the justices of the New York Supreme Court, county judges, district attorneys, and sheriffs. A number of medical men also received the survey.[41] One hundred and ninety-nine questionnaires were returned. Of these, eighty were in favor of retaining hanging and eighty-seven "were either decidedly in favor of electricity or in favor of it if any change was made." Eight preferred poisons, five the guillotine, four the garrote; seven were in favor of other methods, and eight did not express an opinion. The New York Supreme Court justices were equally divided between hanging and electricity. Believing electricity more humane than hanging, one respondent favored "hanging for men and death by electricity for women." Another wanted hanging in aggravated cases, and electrocution when there were mitigating circumstances. Overall, those who opposed change favored electricity if change were deemed necessary. The committee interpreted these results as supportive of a change from hanging to electricity. However, not only was the difference between the supporters of hanging and the advocates for electricity small (eighty-seven to eighty), many of the respondents who were placed in the pro-electricity category were actually in favor of retaining hanging and only supported electricity if a change had to be made.[42]

While most of the survey's respondents merely filled out the questionnaire, one man, Dr. J. Mount Bleyer, submitted an entire report. His interest in execution methods had been stimulated by the governor's message of 1885, and when the Medico-Legal Society—an influential organization of medical and legal professionals devoted to the study of medical jurisprudence—decided to submit a report on the best method of executing criminals, Bleyer was appointed to its committee.[43] His recommendations were not only adopted by the Medico-Legal Society but by the commission itself. Bleyer argued with considerable authority that, by virtue of their extensive knowledge of human physiology, medical men were in the best position to make informed decisions about methods of execution. He began his report with an overview of the problems and benefits associated with hanging. Bleyer noted that hanging was ordinarily quick and painless, but was uniquely open to errors of miscalculation, thus resulting in slow, painful deaths. Such a flawed method of taking human life, according to Bleyer, ought to be replaced in the late nineteenth century.[44]

Electrocution, on the other hand, involved no such prolonged struggle, and Bleyer believed it to be far superior to hanging. He suggested that electric light wires might discharge the criminal, but noted that they would have to be protected from sabotage by the friends and relatives of the condemned. According to Bleyer's plan, a small wooden house, no larger than a watchman's hut, could be built to hold the prisoner for electrocution. He proposed that the criminal stand barefoot on a metal plate connected to one of the wires, and the other wire drop from the roof, just enough to touch the top of the head. The hair should be cut short and moistened, ensuring proper contact. He wrote that the sheriff could merely touch a button, closing the circuit and killing the condemned. Bleyer's plan was tested on a large dog, and, since the dog did not make "so much as a whimper," he was confident his method would kill painlessly.[45]

While Bleyer preferred electrocution to hanging, he also examined a number of other possibilities he believed were superior to hanging. His opinions are notable because they forecast the introduction of lethal injection and the gas chamber. Bleyer claimed that an injection of morphine with a hypodermic needle would produce a painless death. He said the condemned could be executed on his bed in his cell with a 6-gram injection of sulfate of morphine, which would put him to sleep. Bleyer also claimed that within thirty minutes, the condemned man's heart would stop and he would be dead. He deemed this method "certain, painless, undramatic, fault-free, and inexpensive." Bleyer freely acknowledged that some people had doubts about the ability of sheriffs to inject morphine, but said it was a simple technique that anyone could learn quickly. In any case, he claimed, the dying process would be peaceful, as the condemned man would gradually drift off into an irreversible slumber. And, most important, there would be no visible signs of pain. Bleyer also noted that even if the prisoner has been an "opium-eater," the time between his sentence and execution would be enough to clear his system and thus ensure the effectiveness of the lethal injection.[46]

Bleyer further explored the possibility of death by chloroform, a distant precursor to today's gas chamber. Chloroform was a commonly used anesthetic in medicine, and he recommended a large dose of chloroform to stop both the lungs and heart. The criminal might be easily dispatched by placing him in a reclining position and holding a cloth saturated with chloroform over his mouth and nostrils. However, this method might result in a prolonged struggle between the prisoner and his executioners. It also violated the tradition of the condemned's last meal, since an empty stomach was necessary for the chloroform to be effective.[47]

Bleyer acknowledged the validity of the popular argument that criminals dreaded death by hanging more than they would by electricity, morphine, or chloroform, and were thereby deterred from committing crimes. He noted, however, that the injection of a poison deprived the criminal of dramatic displays of courage and bravado. This difference, Bleyer claimed, would compensate for any loss of dread on the part of the condemned man.[48]

While the Gerry Commission gathered information, Dr. Southwick became alarmed that Elbridge Gerry did not share his enthusiasm for electrocution. He wrote to Thomas Edison on November 8, 1887, to ask for help. He told Edison that his "reputation as a scientist and especially as an electrician" prompted him to solicit Edison's opinion on capital punishment and electricity. Specifically, Southwick asked Edison for advice on the strength of current necessary to kill with certainty, and inquired about the likely expense of such an undertaking. Southwick proposed that a chair with wired metal arms would do the job nicely, and he solicited Edison's opinion on the subject.[49]

A month later, Edison wrote back to say that, as a progressive and a free thinker, he was a lifelong opponent of the death penalty, and he did not believe that the law had the right to kill. Implying that he did not want anything to do with Southwick or the Gerry Commission, he expressed the view that life imprisonment was as effective as the death penalty in deterring murder.[50]

Undaunted, Southwick wrote to Edison again on December 5, 1887, this time appealing to the inventor's sense of progress. He claimed that because capital punishment existed and probably always would, "science and civilization demand some more humane method than the rope." Southwick also appealed to Edison's sense of pride, noting that a recommendation from him would carry great weight with the commission. He closed by expressing hope that Edison would change his mind and allow the commission the benefit of his extensive knowledge.[51]

Edison's answer to this second request for help was quite different than his first. Realizing the possibility of using the commission in his campaign against alternating current and George Westinghouse, he replied that he had "carefully considered" Southwick's remarks, and although he would prefer to abolish capital punishment, he recognized the practical necessity of finding the "most humane method available" for executing convicted criminals. He added:

> The best appliance in this connection is, to my mind, the one which will perform its work in the shortest space of time, and inflict the least amount of suffering upon its victim. This, I believe, can be accomplished by the use of electricity, and the most suitable

apparatus for the purpose is that class of dynamo-electric machinery which employs intermittent currents. The most effective of these are known as "alternating machines," manufactured principally in this country by George Westinghouse. . . . The passage of the current from these machines through the human body, even by the slightest contacts, produces instantaneous death.[52]

Southwick was delighted with Edison's second response. He immediately showed it to Elbridge Gerry. Gerry, a great admirer of Edison's, was duly impressed with the letter from the Wizard of Menlo Park. He would later testify that a letter from "the greatest electrician of modern times" was what finally convinced him to recommend death by electricity.[53] Southwick had achieved his goal; Edison's letter was later quoted by the commission in support of its recommendation that hanging be replaced by electrocution.[54]

TWO YEARS after its appointment, on January 17, 1888, the Gerry Commission unveiled its report. The report began by examining the barbaric history of capital punishment. According to the committee, the death penalty had two main purposes: taking the criminal's life as a punishment for his crime, and deterring other "evil-minded and dangerous persons" from committing crimes. The commission noted that in ancient law, numerous offenses were punishable by death as a way of keeping crime in check. The commission cited both the Mosaic code and Athenian law as examples.[55] Furthermore, the report noted, in the late eighteenth century, England punished over one hundred offenses by death, but ultimately discovered that the application of the death penalty to minor crimes did not have a deterrent effect. Any deterrent effect that the fear of hanging may have had was diminished by the indiscriminate use of it for minor offenses. Since the criminal could be executed for stealing a sheep, there was nothing to prevent him from committing a murder if he was caught in the act. The commission reported that, by the late nineteenth century, England had limited the death penalty to six offenses: treason, murder, piracy, military offenses, and crimes involving grave injury to the individual or the community.[56]

The report observed that the infliction of capital punishment was less barbaric in America. When William Penn founded Pennsylvania in 1675, he drew up a system of laws that eliminated the English tradition of capital punishment for petty offenses, rejecting the death penalty for every offense except murder. The commission then criticized earlier methods of capital punishment as unnecessarily cruel and barbaric, claiming that they all rested on an intense desire to exact vengeance and inflict pain and suffering. In the modern era, the commissioners wrote, society had no need to intensify the pain and suffering of the condemned. It needed to dole out punishment humanely and refrain from unnecessary cruelty. Since all existing methods contained elements of "barbarous cruelty," the state of New York needed a modern and humane method of execution.[57]

The commission then examined thirty-four methods of capital punishment. In order to avoid any indication of preference, the commission listed them in alphabetical order. They were:

beating with clubs, beheading, blowing from a cannon, boiling, breaking on the wheel, burning, burying alive, crucifixion, defenestration [throwing out a window], dichotomy—i.e., cutting into two parts, dismemberment, drowning, exposure to wild beasts (especially serpent's fangs), flaying alive, flogging, garrote, guillotine, hanging, hari kari, impalement, iron maiden, peine forte et dure—i.e., placing a heavy weight on the chest, which gradually reduces breathing, poisoning, pounding in mortar (Proverbs 27:22), precipitation—from a lofty precipice, pressing to death, rack, running the gauntlet, shooting, stabbing, stoning or lapidation, strangling and suffocation.[58]

The report then noted that of the thirty-four methods of inflicting the death penalty, only a few were practiced in the "civilized world." A closer look at these methods and the major objections to them lends valuable insight regarding why the commissioners chose to recommend electrocution. The report cited as its primary objection to the guillotine the "profuse effusion of blood" that was "needlessly shocking" to those who must carry out or witness the execution. Even if the killing was done in private, the public would still learn the gory details of the proceedings, which would arouse strong emotions. The report quoted a writer from *Harper's Magazine* who captured antiguillotine sentiment. "The guillotine is apparently the most merciful, but certainly the most terrible to witness, of any form of execution in civilized Europe. The fatal chop, the raw neck, the spouting blood" are demoralizing to witness, as they cannot fail to stimulate a thirst for blood among the people. The commission dutifully noted the special advantages the guillotine possessed; it was quick, pain-less, and certain, for the prisoner could not be resuscitated. Nevertheless, they concluded, its association with the wholesale slaughter during the French Revolution would forever tarnish the guillotine's appeal to the American people.[59]

The garrote had the special benefits of "celerity and certainty," areas in which hanging was deficient. Originally devised by the Moors and Arabs, the garrote utilized a cord, wrapped around the neck and tightened by a wooden stick containing a screw. Garroting employed two techniques for killing: either the condemned died from the destruction of his spinal col-umn by the screw or he died from suffocation by the tightening of the cord. This was generally regarded as its main advantage, but the commis-sion said that to combine two methods in one technique merely multi-plied its barbarity. Although there was little blood spilled, the commission said medical men believed the "fatal screw" was not a reliable means of execution. Spain and her colonies were the only places the garrote was ever employed, and the report noted that although it had been in exis-tence for more than a century, no other nation had ever adopted the method.[60]

Death by firing squad was a method of execution confined mostly to the military, and the commission decided it would be inappropriate else-where. In the military, there were plenty of "competent executioners" to comprise a firing squad. In civilian life, however, a firing squad would likely be bloody and unreliable. The report also deemed it unsatisfactory for it required several executioners, and seemed more suited to a military dictatorship than to a democratic republic. The commission also believed that the firing squad would be disruptive to the orderly administration of justice since it encouraged "the untaught populace to think lightly of the fatal use of firearms."[61]

Death by hanging—America's current method of execution—was given detailed scrutiny by the commission. There were several persistent problems with hanging. First, there was the common practice of providing alcohol to the condemned immediately prior to his execution. Although everyday prisoners were forbidden from drinking alcoholic beverages, the rule was relaxed for those condemned to death. The condemned ordinar-ily took an intoxicating beverage to "stupefy" himself and thereby lessen his perception of the "pain of dying."[62]

Liquor also helped the condemned prisoner muster the fortitude to face his impending ordeal. The Gerry Commission objected to this rationale for supplying prisoners with alcohol; they believed the prisoner should be fully conscious of his imminent death, and under no circumstances be allowed to drink. Drunkenness in itself was a crime, the commissioners pointed out, and it was simply improper to send an intoxicated man to his death. They also claimed that liquor, given to a condemned man to help him face death, might encourage him to resist—or even commit suicide.[63]

After noting that cultured and high-minded people had always held a prejudice against hanging, especially when things went wrong, the commission outlined a series of persistent problems. They ranged from excessive resistance or suffering by the condemned, to brutal indifference by an unskilled or sadistic executioner, to unruly mob behavior. Other drawbacks included difficulties caused by the execution of more than one prisoner at a time, and the sympathies aroused in the public by the killing of a woman or recently disfigured prisoner.[64]

Botched hangings were all too common and thus of great concern to the commission. Poorly run executions created horribly disturbing displays of torture that often led to revolt against the authorities. The end result, claimed the commission, was a profound loss of respect for the law. To emphasize this point, the commission recounted the details of several botched executions from all over the English-speaking world, each more ghastly than the last.[65]

For example, a man named Gow was hanged at London's Newgate Prison and suffered frightfully during his execution, due to some relatively common circumstances. After Gow dropped through the gallows' trapdoor and struggled violently for over ten minutes, his friends, hoping to hasten his death, pulled on his legs with so much force that the hangman's rope broke. Gow dropped to the ground, still alive; he was marched up to the gallows for a second attempt, where he was finally hanged until dead.[66]

Resuscitation was another issue the commission examined. Without taking a position on the likelihood of resuscitating a hanged man, the committee acknowledged the widespread public belief that such a feat was possible and had provided the opportunity for numerous criminals to escape the gallows. The committee recounted the folktale of the near death of Margaret Dickson in the 1730s. The Scottish murderess was convicted of concealment of birth and condemned to death for her crime. After hanging on the gibbet at Edinburgh for almost an hour, her body was cut down and released to her friends, who placed it in a coffin and drove off in an oxcart. During the trip the cart stopped abruptly, loosening the lid of the coffin. Eyewitnesses speculated that the air let in by the unsecured coffin lid and the jolting of the cart revived her. Margaret Dickson lived to have other children; around Edinburgh, she was affectionately known as "half-hanged Maggie."[67]

Concluding their investigation of hanging, the commission stated that it was time for a "radical change." After a "long and exhaustive examination of the subject," they could see no justification for continuing to use hanging as a means of execution. "The deprivation of life is, in itself, the most serious loss which any human being can suffer," the report stated. Hanging inflicted unnecessary pain upon its victims, and would be justifiable only for its deterrent effect—an effect that was never achieved, they noted.[68]

The commission also noted that the infliction of cruel and unusual punishment was prohibited by the New York State Constitution. Although "cruel and unusual" was subject to judicial interpretation, the commissioners judged it safe to assume that punishments involving torture were forbidden by state law. The commission explained that a punishment must be both unusual and cruel to be outlawed; for example, hanging was cruel, but widely practiced and therefore not unusual, while other methods might be unusual, but not cruel. Any punishment that subjected the condemned to additional physical pain was contrary "not only to the humanity of the age, but to the law itself."[69] The commissioners said they were convinced that there was no need to inflict pain in order to satisfy the requirements of the law.

Although death by lethal injection met the commission's criteria that the new method be quick and painless, they decided against its employment. The medical profession was adamantly opposed to the use of the hypodermic needle in executions. They did not want the syringe, which was associated with the alleviation of human suffering, to become an instrument of death. The Gerry Commission further feared that it might prove impossible to find a medical man to act as an executioner. And only a physician could perform the delicate task of injecting a poison such as prussic acid or morphine. Prussic acid was considered too dangerous for an untrained person to handle properly. In apparent contradiction to its own expressed views, the commission opined that the injection of morphine was unacceptable because it would eliminate the "great dread of death" and make executions "too painless."[70]

Finally, the commission turned its attention to death by electricity. It claimed that an electrical execution would be quick, painless, and certain and that it was society's duty to make use of scientific discoveries to further the common good. Arguing for electricity, the commission stated that the condemned would feel no pain because an electric current traveled at the speed of one hundred thousandth of a second, or ten thousand times faster than the nerves could communicate the sensation of pain to the brain.[71]

The committee supported this claim with a long letter from Professor Elihu Thomson, a nationally known electrical expert and cofounder of the Thomson-Houston Electric Light Company, which later merged with the Edison General Electric Company to form General Electric. Thomson began by acknowledging that the amount of electrical current required to produce death depended upon a combination of factors. Most of the time, he contended, death was due to the rupture of blood vessels or injury to the heart. Thomson asserted that alternating currents definitely produced the most injurious effects. He suggested that for somewhere between $100 and $200 a "small machine" could be constructed to execute criminals. The best way to produce death, Thomson continued, was to have the current run from the top of the head to the bottom of the spinal cord. He claimed that this would cause a "painless extinction of all the faculties"; if kept on long enough, it would preclude the possibility of resuscitation.[72] In Professor Thomson's learned opinion, electricity would produce a humane and painless death.

Next, the commission quoted from the letter Thomas Edison sent Elbridge Gerry, testifying to the superiority of alternating current for execution and suggesting the use of a Westinghouse dynamo. Gerry would later testify about how influential this letter was in guiding the commission toward recommending electricity for use in executions.[73]

The commissioners then recounted their experiences as witnesses to a number of experimental electrocutions conducted by Dr. George E. Fell, a Buffalo physician and surgeon who had a macabre interest in executions. The commissioners quoted at length from Dr. Fell's report. In July 1887, Buffalo decided to rid its streets of the many "curs," or stray dogs, roaming the city. The Society for the Prevention of Cruelty to Animals was given the task of exterminating these animals. Dr. Fell, a member of the society and friend of Southwick's, recommended that the job be done with electricity. The killings were conducted at an old police station; the execution apparatus consisted of a "common pine box, lined with zinc." The box was half filled with water, "a muzzle with a copper bit was strapped over the dog's mouth," and the muzzle was connected to an electric streetlight. Once the animal was placed in the box, the switch was turned on, resulting in instantaneous death.[74] Dr. Fell concluded that "electricity will kill quickly." He assured the commission that an "ordinary electric light current . . . is sufficient to cause the instantaneous death of a human being." From his observations, Dr. Fell made the following deductions: "Electricity was the most rapid and humane method of execution, resuscitation was impossible, and the current should pass through the brain."[75]

The commission expressed its confidence that any electrical appliances required for carrying out an execution could easily be manufactured. They deemed a chair with a headrest and footrest essential; one electrode should be connected to the headrest, and the other to the metal footrest. They estimated that the chair would cost about $50. According to the report, if streetlight wires supplied the electric current, it would cost very little to connect the chair to a power source. Depending on its length, an independent power line would cost between $250 and $500. If an independent generator delivered the electrical current, the cost would be between $200 and $500. Maintenance would be minimal. The commissioners felt that the ideal installation would establish independent power sources at the state prisons in Auburn, Sing Sing, and Dannemora. They called for an additional connection to the electric streetlight wires so that a backup source of current would be available if the independent power source failed.[76]

The commission then considered some of the objections to the use of electricity for execution. They acknowledged that various counties might have financial difficulty in acquiring the execution apparatus along with adequately trained personnel. But perhaps the major objection to electricity, the commissioners noted, was its unpredictability. They admitted that too little was known about electricity's effect on the human body. Lightning might strike one person and kill while another might survive, illustrating that similar currents can often produce different results, depending on individual circumstances. The commission believed that some people had a "special power" of repelling electrical discharges. "If lightning does not always kill, surely we cannot expect death to result from artificial electricity."[77]

The commission responded to such concerns by noting that it had carefully considered such objections to electrical execution, but had decided that electricity sometimes failed to kill because the victim did not get the full force of the current. The report claimed that such problems would not occur when the electrical force was applied scientifically to a "vital part of the body." The commission believed that there would be "no practical difficulty" in building an electrical execution apparatus that would take life quickly and painlessly.[78]

The report then quoted thirteen excerpts from their survey, eight of which were from doctors and other medical professionals. The survey's fifth question asked whether the respondents favored delivering the criminal's body to his friends and relatives or allowing medical men to dissect it. The commission agreed with the response of Judge Lucius N. Bangs of Buffalo. "As the life of the criminal is forfeited to the state," he said, "the body should be also." Bangs maintained that the state had the right to use the criminal's body for scientific and medical research. While this was not a part of the punishment, Bangs said there was reason to believe that it may act as a deterrent.[79]

Next, the report quoted Dr. Alfred Ludlow Carroll of New Brighton, who argued that dissection would increase the deterrent effect of capital punishment. "To the most dangerous of the criminal class, the scaffold bears no associations of disgrace; it is rather a heroic culmination of a career of proud and daring defiance of the law." Carroll said most of the criminal class, however, had an extreme aversion to the disfigurement of a dead body and claimed that the dissection of the body in medical schools would be "more dreaded than the death penalty itself."[80]

The discussion of dissection led the commission to its final point, the deterrent effect of capital punishment. It noted that many people remained in favor of hanging because they believe that such a "degrading and revolting system of death" would help to deter potential criminals. The commission wrote that a careful examination of crime statistics, however, did not support that conclusion. For the most part, the members of the commission subscribed to the widely held belief that the criminal classes were extremely superstitious and that they were gravely concerned with the fate of their body after death. According to the commission, one of the reasons hanging seemed to have such a weak deterrent effect was that the body was turned over to friends and relatives for burial and the dead man was regarded as a martyr. In New York City, friends of the dead man, "his companions in crime," often indulged in the most "drunken and beastly orgies." Instead of acting as a deterrent, public hangings tended to celebrate crime and glorify the criminal. The opportunity for the condemned man to display courage and valor by walking steadily from his jail cell to the scaffold only added to his exalted status.

" 'He was game to the last,' has been many a ruffian's eulogy," the commission wrote.[81]

The commission then took up the argument for moving the place of execution from the jail or prison yard of the county in which the offense was committed, to a death chamber deep within prison walls. They suggested that after being sentenced to death, the condemned be taken to a designated state prison to await execution. This would diminish the opportunity for escape and reduce the chance that an angry crowd might attempt a rescue or otherwise breach the rules of public decency.[82]

Clearly unhappy with press accounts of executions, the commission advocated limiting them to a simple statement that the death sentence had been carried out, that justice had been served.[83] The commission claimed that these restrictions were not designed to limit the freedom of the press, but to prevent the creation of a "vicious and morbid appetite" for public spectacles of suffering. This chronicling with "painful fidelity" of the most hideous details of executions had served only to stimulate crime. The commission expressed its confidence that the press would not regard these restrictions as curtailing its liberty, since they were no more severe than the limits that already existed concerning divorce proceedings.[84]

Regarding the carrying out of executions by electricity, the commission made the following recommendations:[85]

> *First.* That the present method of inflicting the death penalty be abolished, and, as a substitute, that a current of electricity, of sufficient intensity to destroy life instantaneously, be passed through the body of the convict.
>
> *Second.* That every such execution take place in a State prison, to be designated by the court in its judgment and death warrant, and that the time of the execution be not fixed by the court, except by designating a period within which it must take place.
>
> *Third.* That immediately after the execution a *post-mortem* examination of the body be made by the physicians present, and the remains be then handed over to the medical profession for further dissection, or be buried without ceremony in the prison cemetery or grave yard, with sufficient quicklime to ensure their immediate consumption.
>
> *Fourth.* That the public accounts of the execution be limited as regards its details.
>
> *Lastly.* Your Commission, pursuant to the direction contained

in the statute creating them, respectfully submit the foregoing to your consideration, and annex hereto, as part of their report, a proposed act, which they believe will render effective the changes they have suggested.

The Gerry Report is considered the founding document for modern execution practice and protocol. The committee's work would have a dramatic impact on the future of the death penalty, not just in New York, but across the nation.

IN THE WINTER OF 1888, Charles T. Saxton, chairman of the assembly's judiciary committee, introduced a bill based on the recommendations of the Gerry Commission. Saxton, a well-respected member of the assembly and Southwick's friend, gave the principal speech in support of the bill. He had anticipated some opposition to "the new-fangled notion" of electricity by those who opposed innovation, but he did not expect it to be so well organized. The Catholic assembly members opposed the section of the bill that mandated that the dead man's remains be buried in the prison yard, without religious rites, and doused with quicklime to hasten decomposition.[86]

At the opening of the session, the floor and galleries of the capitol were filled to capacity with legislators and spectators. The bill to substitute the "mysterious force of electricity for the rope" was introduced, and Mr. Saxton made a passionate plea for its passage. Then came the real struggle; "ridicule and taunts" from the members and observers echoed across the chambers. Assemblymen began insulting each other, and the crowd joined in with its own "vicious sarcasm."[87] Throughout this raucous scene, Saxton stood firm and ready.

Assemblyman Longley, a war veteran representing Brooklyn, told of the pain and agony relatives suffered when they could not bury their dead. He told of the "sad comfort" friends and relatives received from the burial of the bodies of those who had fallen in battle. "I move to amend," proclaimed Mr. Longley, "that friends or relatives may claim the body of the executed man." Although the emotion of the crowd was clearly on the side of the assemblyman from Brooklyn, Saxton did not back down. He told the crowd:

When a man by his crime forfeits his life to the State, the State has undoubted right to dispose of the murderer's body as public policy may direct. What comfort can be afforded those who loved the criminal by viewing the remains which in life had failed of self-respect, and which in death bear the stamp of the State's righteous desecration? In Chicago, the bodies of executed criminals were exposed to relatives and to the public, and that city was brought very close to an insurrection. Public policy would have been better served had the provisions of this bill been operative there.[88]

After a long silence, Saxton faced Longley and looked him straight in the eye. "Finally," he asked his fellow legislator, "does the gentleman wish to force upon me, who served as did he in the Union armies, a comparison of the sacred sorrow for the dead soldier with the passion of regret over the corpse of a dead murderer?"[89] Departing from the usual hoots and howls, the crowd waited in silence, struck by the seriousness of the exchange between the two men.

The silence was broken when Roesch rose to speak for the Catholic opposition. "I hold that where relatives claim the remains, the State has no right to retain them. Property exists in human bodies, and besides, this bill takes away the right of burial in consecrated ground." "Consecrated ground," roared Saxton. "Is the plea here made that hardship follows retention of a criminal corpse from consecrated ground? The criminal who in life would not respect the flesh protected by the law he broke may not in death have demanded for it that which himself had forfeited."[90]

"But," Roesch shouted, "was not the body of Christ stamped with the mark of the law's desecration, and was it not afforded decent burial?" The exchange between the two assemblymen had turned increasingly vehement; the hundreds of people in the chamber stood silent, awaiting Saxton's response. "I will not stand here," he replied, "to answer a question based upon the association of the Holy Savior's memory with that of men executed by the State of New York for murder."[91]

After things calmed down, by a vote of 87 to 8, the assembly passed a bill substituting electricity for hanging in capital cases. The Catholic opposition succeeded in bringing about a compromise; after a postmortem examination of the body by medical men, relatives were allowed to claim the body and have a private religious service within the walls of the prison. If the body were not claimed, it would be buried in quicklime in the graveyard adjacent to the prison without a religious service.[92]

The bill did not pass easily through the senate. Because Saxton's good friend MacMillan had chosen to retire from the senate, Henry Cogge-shall of Waterville, an upstate Republican senator, introduced the electri-cal execution bill. Without notice or comment, the Senate Judiciary Committee removed the central provision of the bill calling for the sub-stitution of electrocution for hanging.[93] Ordinarily, this would have been enough to kill the bill, especially since the session was coming to a rapid close. But on the last day of the session, Senator Coggeshall succeeded in introducing an amendment to restore the electrocution provision, and the restored bill passed on a voice vote. On June 4, 1888, Gover-nor Hill signed the electrical execution bill into law. It took effect on January 1, 1889. Commissioner Matthew Hale, an accomplished legal scholar who had served briefly in the New York Senate, wrote the actual statute.[94]

The law included a strict gag order that prohibited the press from reporting the details of an electrical execution. At the last minute, how-ever, for reasons that remain unknown, but probably having to do with his reluctance to carry out the first electrocution, Warden Durston decided to allow Frank W. Mack of the Associated Press and George G. Bain of the United Press to witness Kemmler's execution. Both were admitted as pri-vate citizens, but with the understanding that they would be allowed to publish their reports.

The gag order, while almost unilaterally ignored, was subject to much acrimony at the hands of the press. On its editorial page, the *New York Times* attacked the restrictions on the freedom of the press. It claimed that it had no objection to the reasons behind the ban, but argued that requiring good journalistic practices was not the proper domain of crimi-nal law. It gave three main reasons: first, a total ban on all news reports was too "indiscriminate." Certain executions, like that of President James Garfield's assassin, Charles Guiteau, which took place in 1882, are "important historical events." Besides, it was impossible to develop a steadfast rule to determine which executions could be covered by the press. Second, the criminal law should not attempt to legislate "decency" or "good taste." The *Times* argued that if a newspaper played to "depraved taste," its reputation would suffer and its readership decline. Third, the *Times* editors wrote, an a priori ban on newspaper accounts of executions would be entirely unenforceable.[95]

The *New York Evening Post* was one of the only newspapers to endorse wholeheartedly the new bill, claiming that it would put an end to the bar-barism of hanging. The paper argued that little positive could be said for hanging; it occasionally inflicted "torture on the culprit" due to his strong constitution or the hangman's bungling. The *Post* wrote that its original position in English jurisprudence was to warn potential criminals of their potential fate. Its editors believed there was no reason to preserve this "cruel absurdity" in late-nineteenth-century America and that the horror of execution day had been heightened by the practice of regarding the criminal as a celebrity or hero. The criminal's cell was traditionally open to the public; reporters interviewed him incessantly, seeking out "maudlin reflections" on his grave situation. Thankfully, according to the *Post*, the new law would put an end to this. After the sentence was pronounced, the condemned would be kept in near complete isolation. The day of his execution would not be announced to the public, and his death would be attended by just a few witnesses. If his body were claimed by his relatives, they would not be allowed to have a public wake or funeral. The *Post* applauded these restrictions, believing they would mend New York's system of capital punishment.[96]

The *Evening Post* responded to the *Times*'s biting editorial by endorsing the prohibition on press reports and urging the arrest and prosecution of any editor who violated the law. The *Times* ridiculed the *Post*'s position, pointing out that it too had violated the law by publishing "details" of Kemmler's execution, and it predicted that the *Post* would continue to violate the law with detailed stories of the execution.[97] The *Sun*, a popular paper with a large circulation, agreed with the *Times*. After the execution it argued that the publication of the awful torment and execution of Kemmler performed a valuable service. And it pointed out that if the law had not been violated, the public might never have learned of the "horrible circumstances" of Kemmler's death.[98]

As one might imagine, the *Post* and its editor, E. L. Godkin,[99] received little support from other publications. The *New York Press* observed that electrical execution was supposed to be quick and painless, and that the maudlin and sensational features of hanging were to be replaced by a less dramatic, more scientific, and, most of all, humane method of execution. To this end, the Electrical Execution Act prohibited newspapers from publishing the details of an execution.[100] The intention of the law, wrote the *Buffalo Express*, was discredited by the outcome of the first electrical execution. Had not the press openly defied the gag law, the public would not have been aware of the terrible scene that took place at Auburn.[101] The *Buffalo Courier* supported this position, declaring that the law failed to eliminate the "revolting and horrifying features" of judicial execution.[102]

The Medico-Legal Society of New York, however, enthusiastically endorsed the restriction on press reports of executions. They argued that newspaper people lacked the medical knowledge and scientific sophistication to interpret correctly the physiological reactions of the condemned man to the deadly current. The *Albany Law Journal*, closely aligned with the New York bar, complained that newspapers ran "false and sensational accounts" of Kemmler's execution both to seek "revenge" for their exclusion from the death chamber and to sell newspapers. Despite the blatant violation of the law, the state never attempted to prosecute Warden Durston or any of the offending reporters, editors, or news organizations.[103]

WITH THE PASSAGE of the new Electrical Execution Act, all the elements of the modern execution were put in place. The act did more than change the method of execution from hanging to electricity. It helped establish an elaborate set of rules and procedures that still delineates the protocol of modern executions, whether the killing is accomplished by electricity, gas, firing squad, or lethal injection. Most important, the act helped change the social meaning of executions from a public ceremony orchestrated to venerate the power of the law and the primacy of the soul, to an empty private ritual calculated to produce death on a fixed schedule.[104]

During the period after the 1835 act, which prohibited public executions, and prior to Kemmler's execution, not much had changed in regard to how prisoners were put to death. Despite the fact that executions were supposed to be private, they remained large and lavish spectacles of suffering. It was not until 1890 and Kemmler's execution that the measures from the 1835 act were effectively enforced. Never again would large crowds of spectators line the streets and rooftops of New York's cities and towns to witness the condemned criminal's "death march" up to the gallows, or picnic as the noose was tied around his neck. With the introduction of the Electrical Execution Act of 1888, the practical and social meaning of executions had forever changed. To understand just how fundamentally execution etiquette was altered, it is instructive to contrast Kemmler's execution with a typical high-profile execution just ten years earlier—that of Samuel Steenburgh.

Steenburgh, a Negro laborer accused of murdering his drinking companion, a white farmer named Jacob Parker, was hanged in his hometown of Fonda, New York, on April 19, 1878. Only four hundred people were supposed to be allowed to attend Steenburgh's "private" execution, even though the "good-natured Sheriff, Stephen Fonda," for whose family the town was named, has been "literally besieged for weeks" for tickets to witness the execution. Sheriff Fonda was forced "to refuse a great majority of these morbid curiosity-hunters," several of whom asked for passes for their wives or girlfriends. Three armed companies of militia were employed to keep order, and extra trains were run to transport spectators from as far away as New York City. The execution was set to take place between the hours of noon and 1 p.m., "to enable visitors to return to their homes on the afternoon trains."[105] On the day of the hanging, a swarm of more than fifteen thousand people descended on the tiny village of Fonda. A headline in the *New York Times* read, "Fifteen Thousand People Make a Picnic of the Hanging . . . He dies amid the jeers of the mob."[106]

William Kemmler met his fate before dawn deep within the bowels of the state prison at Auburn. Only a small band of professionals was present. None of them were curiosity seekers or bloodthirsty drunken revelers bent on having a good time. All were present as part of their duty to law, science, or medicine. They wore solemn, grave expressions on their faces. Their manner was polite and respectful. Some appeared as if ashamed of their participation, having deliberately intruded upon Kemmler's private moment of anguish. Others projected a genuine sense of satisfaction or even pride, for they honestly regarded the first electrocution as a great humanitarian achievement. None saw Kemmler's early-morning execution as an opportunity to frolic or sneer at the condemned.

For much of the nineteenth century, the period between conviction and execution was less than a week. Prisoners condemned on a Tuesday or Wednesday might be executed the following Monday—hanging day. There were few appeals of sentences and almost no stays of execution. By the end of the century, however, the time spent between sentencing and execution had expanded to a few months. Steenburgh spent five months in confinement while his lawyer pursued a new trial based on fresh evidence of his innocence. But when additional evidence of Steenburgh's guilt emerged, his lawyer was forced to give up his appeal. Kemmler, on the other hand, was granted two stays, permitting his attorneys enough time to pursue appeals. In all, fifteen months elapsed between Kemmler's conviction and his execution. His five appeals went through three different levels, from the Cayuga County Court to the New York Supreme Court to the U.S. Supreme Court.

Prior to the Electrical Execution Act, the condemned was held at a local jail in the county where his crime took place. He was usually permitted to entertain visitors and members of the opposite sex. To some extent, he was treated like a celebrity. Steenburgh, for example, was held at the Fonda jail, a small stone building lacking an enclosure. There he entertained guests, charging each ten cents for the privilege of visiting him.[107] Steenburgh even made a deal with Sheriff Fonda and County Clerk Burtch. The two men paid him $100 for his twenty-two-page confession detailing seven murders and hinting at several others, which they put on sale for twenty-five cents, selling all five thousand copies within a few hours. In addition, Steenburgh sold his body to five physicians from the Albany Medical Institute for $100, leaving the money to his eleven-year-old daughter, Susie.[108]

From the time of his arrest until four days after he was sentenced to death, Kemmler was detained in the Erie County jail in Buffalo. Then, in accordance with the new law, he was transferred halfway across the state to Auburn where he was placed in solitary confinement. This measure was designed to lessen the possibility of escape or rescue and to make it difficult for the condemned's friends to protest his impending execution.[109] Kemmler was not allowed visitors, except for his lawyer, clergyman, and close family. Without a court order, the public and newspaper reporters were denied access to him. The warden's wife, Mrs. Charles Durston, was his only regular visitor.

Samuel Steenburgh's execution was filled with all the splendor that usually accompanied public hangings. For Steenburgh, drums beat mournfully and priests chanted the Fifteenth Psalm and the "Miserere" as he marched to the gallows. In the background, a violinist played "The Sweet By-and-By." Oddly enough, Steenburgh himself "seemed entirely unconcerned. He walked with a firm tread, and occupied himself in casting curious glances at the scaffold and at the crowd." Once Steenburgh had ascended the gallows, Sheriff Fonda put on his official hat and adjusted his red silk sash. He would act as the executioner. As the sheriff tried to read the death warrant above shouts from the crowd, a priest handed Steenburgh a brass and ebony crucifix, which the condemned man kissed "fervently."[110] When asked if he had anything to say, Steenburgh admitted his guilt and announced that he was prepared to die. The sheriff then placed a black hood on his head. Steenburgh complained that the noose was too tight and Sheriff Fonda removed the hood. After about ten minutes of exchanging quips with people he recognized in the crowd, asking for forgiveness, and pleading for more time to listen to the choir sing, Samuel Steenburgh lifted his head up high and stepped resolutely into eternity. William Kemmler's execution was different. Gone was the public "theater of death," and along with it much of the social and symbolic nature of capital punishment. In its place was a private, impersonal, and highly rational machinery of death, one that emphasized instrumental rather than moral values. Prior to the advent of the electric chair, the executioner was a public figure well known to the community. In Steenburgh's case, he was the town's sheriff. As far as we know, Kemmler never met his executioner; he was an ordinary electrician by trade whose identity was never officially released to the public. Those who watched Kemmler die were professionals, more interested in the science and mechanics of killing than in the meaning of justice. William Kemmler, the man, his crime, his redemption or lack thereof, and the fate of his soul—all were pushed to the background. It was the chair—not the condemned—that took center stage.

CHAPTER FOUR

Harold Brown and the "Executioner's Current"

T HE BATTLE OF THE CURRENTS was in full rage when Harold Brown, a thirty-one-year-old self-taught engineer, entered the fray.[1] Three recent accidental deaths involving arc streetlights powered by alternating current in New York City provided Brown with an opportunity to inject himself into the debate regarding the relative safety of direct versus alternating current. On June 5, 1888, the day after Governor Hill signed the electrical execution bill into law, Brown published an angry letter to the editor in the *New York Evening Post*. In "Death in the Wires," Brown complained that after each accidental death, electrical engineers connected with the company at fault remained silent, while the New York newspapers mistakenly clamored to place all electrical wires underground. Although he acknowledged that underground wires would reduce the risk of accidental deaths, Brown claimed that it was only a half measure. New Yorkers needed to recognize that alternating current was simply too dangerous for residential or commercial use.[2]

In his letter, Brown deliberately aligned himself with the interests of the Edison Company when he suggested that the enforcement of "a few common sense regulations" would make it nearly impossible for arc lights to cause fatalities. And he positively endeared himself to Thomas Edison when he singled out Westinghouse for special condemnation. At the "risk of offending" electric companies who had invested heavily in "unsafe" delivery systems, Brown listed his solutions to the problem of accidental AC electrocution, claiming that no "disinterested electrician" could dispute them. He asserted that direct or continuous current of low voltage, like that used by the Edison Company, was safe. Moreover, Brown explained, the mere fact that a current was of low voltage did not guarantee its safety. He maintained that low voltage must also be direct or continuous, since alternating current, even at low voltages, was always dangerous. In alternating current, Brown wrote, "impulses are given first in one direction, then in the other several thousand times a minute." Brown claimed that it was the rapid reversal of impulse current, not the current itself, that was deadly. A steady current of equal intensity would do little damage.[3]

Brown tempered his argument by acknowledging that AC was not always dangerous any more than DC was always safe, but he noted that only four deaths had occurred with direct-current systems, whereas scores of deaths had occurred as a result of alternating-current systems. Brown granted that if alternating current were perfectly insulated, there was no danger. But he contended that it was an open secret among electricians that the wire known as underwriter's wire used for alternating current was of exceptionally poor quality and could be destroyed in a heavy rain. Among electricians, Brown noted, it was often referred to as undertaker's wire.[4]

Ultimately, Brown argued that even if direct current was not always safe, it was certainly safer than alternating current; he went so far as to label AC as "damnable."[5] He claimed that the only factor supporting the use of alternating current was that it saved money on copper wire. Conveniently, he neglected to inform his readers that this savings was considerable; the smaller copper wire required for alternating current saved investors about $1,000 per mile in comparison to the thick wire necessary for direct current.[6] According to Brown, money-hungry electrical manufacturers of AC exposed the public "to constant danger from sudden death," so that a corporation like Westinghouse could pay its stockholders large dividends. He concluded by suggesting that the New York Board of Electrical Control follow Chicago's lead and ban the use of alternating current altogether. Brown claimed that the use of underwriter's wire with an alternating current was "as dangerous as a burning candle in a [gun] powder factory." Satisfied that he had made his points, Brown then listed six safety rules that New York City must take into consideration; the chief among them was that no alternating current with a higher electromotive force than 300 volts should be used.[7]

Brown's allegations against the Westinghouse Company embroiled the entire profession of electrical engineers in a dispute over the relative safety of alternating and direct current. No longer was this just a battle between Edison and Westinghouse for control of the electrical power industry. Increasingly, New York City's Board of Electrical Control came under pressure from the public to take action to prevent future accidental deaths. As expected, a heated discussion of Brown's allegations dominated the early moments of the board's June 8, 1888, meeting. Compelled to act, Commissioner Daniel L. Gibbens read aloud Brown's entire letter, which must have taken more than an hour. At the conclusion of his reading, the commissioner moved that Brown's letter and proposed rules be published in the minutes of the meeting and that copies be sent to the various electric lighting companies in New York City, most especially to George Westinghouse. Gibbens also suggested that the recipients' opinions be solicited.[8]

In the course of just three days, the little-known Harold Brown had placed himself in the middle of a fierce struggle between two great electric companies for control of an emerging power industry. The common interest of Edison and Brown—one a distinguished inventor/entrepreneur struggling to save his dominant share of the industry he created, the other a young, aspiring electrical engineer trying to make a name for himself—would influence public policy regarding capital punishment in a way that no one could have predicted. With Edison's help and guidance, Brown would affect not only William Kemmler, but the life and death of everyone on death row. For the entire twentieth century, the electric chair would become synonymous with capital punishment in America.

While the controversy raged, Harold Brown approached Thomas Edison to ask for the use of his laboratory to demonstrate that alternating current was more deadly than direct current.[9] Edison recognized how he could use Brown to discredit alternating current and received Brown with great enthusiasm, assigning his chief electrician, Arthur Kennelly,[10] to work with Brown.[11] Much to Brown's delight, Edison himself promised to take a special interest in his work. Indeed, Edison wasted no time in writing to Henry Bergh of the American Society for the Prevention of Cruelty to Animals, asking for some "good-sized" dogs on which to experiment.[12]

The next meeting of the electrical control board was held on July 16, 1888, in the lobby of Wallack's Theater. Some twenty-five people attended and heard several papers denouncing Brown's proposals. T. Carpenter Smith, an eminent Philadelphia electrician, was the first to have his paper

(National Cyclopaedia of American Biography)

Harold Pitney Brown

read. He contended sarcastically that Brown's letter contained "sweeping assertions" that no "practical electrician" would endorse. He wrote that Brown's claim that " 'no disinterested electrician' would deny his statements, is very correct, since . . . some of his statements are hardly worth denying."[13]

Smith went on to charge that Brown's statements were published in the interest of the Edison Company. He pointed out that while Brown was correct in claiming that continuous current had caused only four deaths, and alternating current had claimed scores of victims, Brown failed to note that alternating current had tens of thousands of arc lamps in its system, whereas the continuous current had merely thousands. Smith likened Brown's deceptive exposition to the notion that because only four men had been killed recently while traveling on horseback in New York City and scores had died while traveling on the railroad, it was safer to ride a horse than a train.[14]

Moreover, Smith claimed that Brown lacked experience with alternating current, and said he seemed to have gotten his information chiefly

from the Edison Company's notoriously alarmist "Warning Pamphlet."
Brown referred to himself as an electrical engineer, Smith said, "but what
was his education on the subject?" He remarked that Brown had invented
the "Brown converter," an appliance that converted high-voltage direct
current for arc lighting into low-voltage direct current suitable for incandescent lighting. For two or three years prior to his *New York Evening Post*
letter, Smith noted, Brown tried to convince arc lighting companies to
supply 3,000 volts for incandescent lighting to houses and stores. Clearly,
Smith argued, Brown's new position on the so-called danger of high-
voltage alternating current was dictated by its damage to his financial
interest.[15]

Several other prominent electricians read papers critical of Brown. A
paper from S. C. Peck of New York City pointed out that Brown's concerns about the high-voltage alternating current wired into residences
were unfounded. Peck noted that a transformer outside homes converted
the wires' current of 1,000 volts to only 50 volts, making the voltage safe for
private residences. Peck went on to charge that Brown was masquerading
as a philanthropic electrician; he was really trying to promote the interest
of one electrical company over another.[16]

After papers by several other distinguished electricians were read, a letter from H. M. Byllesby, vice president and general manager of the Westinghouse Electric Company, was presented. Byllesby noted that almost
two years earlier, the Westinghouse Electric Company initiated the commercial introduction of the alternating system of electric lighting. He
claimed that the advantages of the system were immediately apparent to
Westinghouse's customers. The first system was installed in Buffalo, New
York, on November 30, 1886. Less than two years later, ninety-eight stations were in operation and twenty-nine more were under construction.
Byllesby said that until recently it had not been necessary for his company
to respond, but the Edison Company's attacks were now reaching a more
intense level. The Westinghouse representative labeled the Edison Company's attacks "unmanly, discreditable, and untruthful."[17] Furthermore,
Byllesby claimed that alternating current was less dangerous—less likely
to cause fire or death—than the continuous or direct current. He pointed
out that when a direct current broke, it produced a "vicious spark" that
often melted the edges of the switch. With the alternating current,
Byllesby explained, there was very little spark and no heat. He also noted
that there had not been a single fire in the 98 Westinghouse stations then
in operation, while there had been numerous fires at the 125 direct-

current stations, and, at the most recent fire, the entire central station for Boston was destroyed. In closing, Byllesby referred to the Edison Company's warning pamphlet, which he said was a clear indication of how desperate Thomas Edison and his colleagues had become.[18]

Although it was late in the night before all the papers were read, at the board's request, T. Carpenter Smith explained the workings of the underground wiring system used by the Westinghouse Company in Pittsburgh. Smith dissected a converter, demonstrating that the statements made by Brown regarding insufficient insulation were mistaken. President Jacob Hess then invited Brown, or anyone else who wished to argue against the alternating-current system, to step forward. No one did—Brown was out of town—and the meeting was adjourned.[19]

With the tide of the electrical profession turned against him, Harold Brown decided he needed a dramatic demonstration of the dangers of alternating current. On July 30, 1888, by invitation, seventy-five electricians assembled at the Columbia College School of Mines in the private laboratory of Professor Thomas Chandler. They had come to observe Brown's experiments on the effects of electricity on animals. Brown addressed the audience by claiming that he was there on behalf of the many workmen who had been accidentally electrocuted. Not only had they been robbed of their lives, Brown asserted, but their memory had been denigrated by their employers' accusations that their own "foolhardy carelessness" had caused their deaths.[20]

Brown said that even though he realized that by publishing his letter in the *Evening Post* he would invite a barrage of criticism from the "wealthy and powerful" alternating-current corporations, he felt it was his obligation to tell the public the truth. As evidence that these corporations are so powerful, Brown cited the fact that the editor of one newspaper only agreed to print his article if he would omit the sections that criticized alternating current. Brown refused. After two more deaths, he took his article to the *Post*, where it was immediately published. Brown explained that he had been in Virginia, unable to defend himself at the July 16 meeting of the Board of Electrical Control. He acknowledged that in order to convince his critics, he needed to conduct a public demonstration of the lethal nature of alternating current.[21]

Brown told his Columbia audience that there was not much utility in presenting a scientific paper to this gathering; instead, he said, he would present a few samples of the experiments he had been conducting at the Edison laboratories to determine the "death points" of continuous and

(New York Medico-Legal Journal, 1888)

Harold Brown electrocuting a horse at the Edison laboratory

alternating currents. To assure the audience that his experiments would be honestly performed, Brown had invited Arthur E. Kennelly, Edison's chief electrician; Dr. Schuyler S. Wheeler of the Board of Electrical Control; Dr. Frederick Peterson, a physician who specialized in the medical uses of electricity; and one of his major critics, T. Carpenter Smith, to assist in the demonstration. Brown began his experiments by dragging a seventy-six-pound part-Newfoundland dog named Dash onto the stage and into a wooden cage with heavy copper wire woven through the bars. Sensing their discomfort, Brown assured the audience that although the dog appeared friendly, he was actually a "desperate cur" who had already bitten two people.[22]

Dash was muzzled and tied down, and the electrical contacts were made to fit the right foreleg and left hind leg. They were wrapped with wet cotton wadding and tied down with uninsulated #20 copper wire. The animal's resistance registered 15,300 ohms. Brown announced that he would first try the continuous current at a force of 300 volts. When the current was turned on, the dog yelped and then whimpered. After it was shut off, Dash immediately received another 400 volts; this time, he struggled desperately, still yelping. At 700 volts, he tore off his muzzle and nearly escaped from his cage. After a short delay, in which the dog was tied down again, 1,000 volts of direct current were sent through his body. His resistance to the electric current had dropped to about 2,500 ohms.[23]

The audience grew increasingly restless at the sight of such brutal treatment of an innocent animal. At this point, Brown told the gathering, "he will have less trouble when we try the alternating current . . . [it] will make him feel better."[24] A gentleman in the audience demanded that "in the name of humanity," the dog be killed at once. Claiming to be a humanitarian himself, Brown agreed, and a force of 330 volts of alternating current ended the "death struggles" of the tortured animal. A reporter from the *World* demanded an end to this "inhuman performance." At this point, Mr. Hankinson, superintendent of the Society for the Prevention of Cruelty to Animals, showed his badge and ordered the experiments stopped. Experiments of this nature, he said, conducted by "colleges and institutions for the benefit of science" were permissible, but they were not permissible when conducted in the "interest of rival inventors."[25]

Brown lamented the stoppage of his experiments, saying that he had enough dogs to satisfy even the most skeptical of observers. "The only places where an alternating current ought to be used were the dog pound, the slaughter house, and the state prison."[26]

Brown's first experiments at Columbia College School of Mines were generally regarded as inhumane and recklessly biased against the manufacturers of alternating current. In response to this criticism, Brown resumed his experiments on Friday, August 3, under the impartial direction of Cyrus Edson of the Board of Health and Dr. Charles F. Roberts, an assistant professor of physiology at Bellevue Hospital Medical College.[27] A crowd of nearly eight hundred was in attendance.[28] No one from the Westinghouse Company was invited. This time, Brown used only alternating current to electrocute three dogs. The first dog, whose resistance measured 14,000 ohms, was killed by the force of 272 volts; the current lasted five seconds. The second dog had a resistance of 8,000 ohms and he was dead after experiencing the current for five seconds at 340 volts.[29]

The third dog, a "big black beast" with fine, long hair, probably a mix of Irish Setter and Newfoundland, had a "bright, intelligent" look and appeared friendly. His natural resistance was 30,000 ohms. Brown started his deadly experiment with an alternating current of 220 volts for five seconds. Like the other dogs, this beast was paralyzed while the current was on. As soon as the current was off, however, he struggled in great agony.

Even after thirty seconds he remained conscious, and at forty-five seconds he had full respiration, although he was breathing heavily. At sixty seconds, he was still breathing and appeared to be recovering. After four minutes, Brown turned the voltage up to 234 volts for another thirty seconds. The dog stood, paralyzed, for about thirty seconds. Then he collapsed, unconscious, on the floor of the cage. The current was shut off, but the dog's heart continued to beat for another two minutes. Incredibly, it was the opinion of Harold Brown and the other doctors who were present that like the other dogs, this dog felt no pain. Death usually causes the relaxation of the sphincters of the bladder and rectum, the result being a discharge of urine and feces; this dog was no exception. The stench was unbearable.[30]

After the second round of experiments, decisive and determined criticisms of Brown's argument continued. Perhaps the most damaging came from Dr. P. H. Van der Weyde, who had attended both the meeting of the New York Board of Electrical Control and the demonstrations at Columbia College. He presented his observations on Brown's experiments in a speech to the National Electric Light Association, an organization dominated by the alternating-current companies.[31] Van der Weyde began his critique by distancing himself from the hurtful personal attacks made on Brown since his June 5 article. Brown's concern about alternating current had made him the object of near-universal derision and, according to Van der Weyde, this was truly unfortunate.[32]

That said, Van der Weyde observed that Harold Brown's comments and experiments seemed more aimed at demonstrating the superiority of direct current than seeking the truth. For example, when Superintendent Harkinson of the Humane Society stopped the demonstration, Brown was compelled to give the animal the coup de grâce by executing him with an alternating current of only 300 volts. Although Brown claimed this proved that alternating current was deadly, Van der Weyde said it actually proved nothing; the animal was already half-dead from the earlier direct currents of 300, 600, and 900 volts, respectively. Van der Weyde maintained that another dose of 300 or 600 volts of direct current probably would have ended his life just as quickly as the jolt of alternating current.[33]

Next, Van der Weyde made his most telling criticism of Brown. He wrote that, as any electrical engineer ought to know, voltage by itself means very little. The danger comes only when an electromotor force, measured in volts, is accompanied by a sufficient quantity of current, measured in amperes. An electromotive force carrying either a direct or alternating current of only a few "micro-amperes" may be as high as 1,000 volts without causing any damage to its victim. In such a case, the current acts like water leaking out of a pinhole under high pressure. On the other hand, a large quantity of current under low pressure, or voltage, will not even have the power to penetrate the human skin. A small amount of current, no matter how much electromotor force is behind it, is not dangerous.[34]

Van der Weyde further noted that none of the dynamos used in Brown's experiment belonged to Columbia College, but were instead provided by Thomas Edison. He accused Brown of deception, stating that the voltage of the alternating current used must have been double that of the direct current. He gave as evidence the fact that two steam engines and two dynamos were employed to produce the alternating current, whereas, for the direct current, only one steam engine and one dynamo were used. Van der Weyde ended by comparing those who predicted disaster from the introduction of AC to the people who warned against the introduction of illuminating gas, electric lights, and railroad trains. "These alarmists told us that gas would blow up our cities, or destroy them by universal conflagrations, and the railroads would cause the . . . wholesale slaughter [of people] who dared to tempt Providence" by traveling at the outrageous speed of twenty miles an hour.[35]

MEANWHILE, the Edison Company was growing increasingly desperate. During the previous year, it sold forty-four thousand lights to be powered by central stations. The Westinghouse Electric Company received orders for forty-eight thousand lights powered by central stations during the month of October 1888 alone. Since the Westinghouse Company had begun its operation two and a half years earlier, it had sold more central station plants employing alternating current than all the other direct-current companies combined.[36]

In the autumn of 1888, Charles Batchelor, one of Edison's top assistants at Llewellyn Park, was almost electrocuted while performing animal experiments. Although Batchelor was working with direct current at the time of the accident, he told a newspaper reporter that he owed his life to the fact that he was not working with alternating current. Edison used this occurrence to boost his campaign against alternating current. This incident brought reporters to the Edison laboratories, where Edison was asked repeatedly about how criminals could best be executed under the new execution law. "Hire them out as lineman to some of the New York electric light companies," was his ready answer. When asked again, Edison spoke in a serious tone. "There is no reason why there should be any failure in an execution by means of electricity. Its use for that purpose will not require the invention of any new machine or the application of any principle not well understood. Electricity of a high tension must be used, and an alternating one rather than a straight one."[37]

When asked how electricity killed, Edison admitted that he did not fully know the answer. "I do not think that the electricity kills the man direct," he responded. He said it did not strike vital organs and destroy them; rather, Edison explained, it excited the physical forces of nerve and muscle to a fatal pitch. When asked if he approved of execution by electricity, Edison responded in a disingenuous fashion. "I do not approve of any execution," Edison said solemnly. "I think that the killing of a human being is an act of foolish barbarity. Society must protect itself," he acknowledged, but added that it did not need the death penalty to do so. Edison said he much preferred lifelong confinement for criminals; he reasoned that the criminals could be put to productive labor, and he mentioned the possibility of employing them to beautify public buildings as an example. Ever mindful of his public image, Edison added that this would have to be done without "infringing upon the rights of honest laborers." Repeating that he was against the death penalty, he assured reporters that electricity would do the work it needed to, and said it was more certain and perhaps more civilized than the rope.[38]

AFTER THE GERRY COMMISSION'S REPORT was passed into law, the New York state legislature asked the Medico-Legal Society to work out the details of electrical executions.[39] At the September 1888 meeting of the society, a committee was formed to recommend how to put the new law into effect. Dr. Frederick Peterson, who used electricity in his medical practice, was appointed to head up the effort.[40] Dr. Peterson had been Harold Brown's assistant at the Columbia College experiments, and, naturally enough, enlisted Brown's help.

Peterson's committee moved quickly, presenting a preliminary report at the society's November 14, 1888, meeting. The committee tentatively recommended that an electrical current be passed from the top of the head through the spinal column to the small of the back. This would cause immediate unconsciousness. They also suggested that a metal helmet, secured to either a table or the back of a chair, be used to carry 3,000 volts of electricity, either direct or alternating current, but preferably alternating, for thirty seconds.[41]

Because the heaviest of the dogs killed the prior summer weighed only 90 pounds, committee member Professor R. Ogden Doremus inquired about the electromotor force necessary to kill a human being. To this end, Brown conducted a few more experiments at Edison's West Orange laboratory, executing three farm animals. The first was a calf weighing 124½ pounds, the second a 145-pound calf. Finally, to demonstrate that electrocution would be certain and instantaneous, Brown killed a horse weighing more than 1,200 pounds. According to Brown, these experiments demonstrated that alternating current was superior to direct current in its life-destroying properties. Dr. Peterson and Arthur Kennelly assisted Brown; among those observing were Thomas Edison and Elbridge T. Gerry.[42]

WHEN THE NEW YORK LEGISLATURE passed the electrocution bill, it failed to specify the type of current, how much electromotor force would be needed, how much current would be passed through the prisoner, and for how long. It did not even suggest what the execution apparatus should look like. Brown and Peterson would have liked to have had the use of alternating current written into the law, but Dr. Southwick and his commission members felt that it was unwise to bring an amendment before the legislature and risk the repeal of the entire electrical execution bill. Besides, it would prevent the state from executing anyone until the amendment was passed by both the assembly and senate and signed into law by the governor.[43]

It was clear that much work needed to be done before the first execution could go forward. Dr. Frederick Peterson considered most of the prior killing experiments unscientific, and therefore worthless. Most of the dogs killed had been chloroformed first; no measurements were taken, and very few details revealed. As a result, Harold Brown, Arthur E. Kennelly, and Dr. Peterson conducted twenty-seven experiments at the Edison laboratory. They found that the skin and hair had high resistance, and that the reason the dogs varied so much in resistance, from 3,600 to 27,500 ohms, had to do with the thickness and moisture of the skin, plus the density of the hair. These experiments revealed that there was a tremendous variance in the amount of voltage necessary to kill, but that alternating current was clearly superior for this purpose.[44]

Furthermore, the dogs were executed by attaching a wire to a muzzle inside their mouth and having them stand in water, which was clearly not practical for condemned criminals. The Peterson Committee then made some suggestions concerning the size and location of the electrodes and how long to keep the current flowing to produce certain death. Dr. Peterson closed his report by admitting that there were still many unanswered questions; he predicted there would be some "bungling" at the first execution.[45]

At their December 12, 1888, meeting, the Medico-Legal Society endorsed alternating current as the best means of executing criminals by electricity.[46] Of the few changes in the draft from the first meeting, this was the most important. Apparently, Harold Brown's most recent experiments had convinced Dr. Peterson and the other members of the committee that the killing power of alternating current far exceeded that of direct current. Many others, however, were not as convinced, and a heated discussion ensued. Dr. Peterson, who was the society's chief proponent of alternating current, had to withstand a series of hostile questions before the society approved the report as presented.[47] Alternating current was to become the executioner's current.[48] Professor C. F. Brackett of Princeton College denounced the society's decision; he was quoted as saying that New York would never have the "nerve" to perpetuate such an "outrage."[49]

Within days, George Westinghouse broke his long silence by placing an advertisement in all of the major New York newspapers. In it, he argued that high-tension alternating current was less dangerous than direct current. He accused Harold Brown of being paid by and working for the interest of the Edison Electric Light Company. According to Westinghouse, it was no secret that the success of alternating current was destroying Edison's business, and this explained why the Edison people were in a panic. Westinghouse claimed that Brown's experiments were designed to produce "the most startling effects" with the least possible amount of current, and that Brown had manipulated the experiment. Westinghouse further charged that the animal experiments had little to do with science; instead, he said, they were designed to scare the public into thinking that alternating current was too dangerous for residential or commercial use.[50]

Brown responded immediately. In order to make his reply appear to be a letter to the editor, Brown paid an extra fee of $275 apiece to the *New York Herald*, *World*, *Sun*, and *Tribune*[51] for his ad to appear without the identification tag ADV. Brown's December 18, 1888, reply to Westinghouse denied that he was working for Thomas Edison or any of the Edison companies. He claimed that Westinghouse had failed to prove his assertion that alternating current was less dangerous than continuous current. He reminded the public that he had proven his position during two experiments at Columbia College, and argued that his interest was for the safety of the public, whereas Westinghouse was simply interested in making money. Brown ended his response with a mischievous but apparently serious challenge to Westinghouse to participate in an electrical duel. Westinghouse was to be wired to an alternating-current dynamo, while Brown would be wired to a direct-current one. Brown stipulated that the alternating current be at least 300 reversals per second; he proposed that the duel would begin with a jolt of 100 volts and would be increased by 50 volts per round. The duel would end when one of them "publicly admits his error."[52]

Westinghouse ignored Brown's challenge, causing Brown to later allege that Westinghouse "did not dare to risk [his] life to prove the sincerity of [his] statement." Brown further claimed that although Westinghouse lived in an area served by an alternating-current station, his own house used direct current. Brown then issued another, less hazardous challenge to Westinghouse, asking that he submit samples of his equipment alongside submissions of products using DC current for testing at the impartial Electrical Testing Bureau at Johns Hopkins University. Brown's express purpose was ascertaining whether or not AC really was more efficient than DC.[53]

Although Brown had succeeded in getting New York State to embrace alternating current as the best means of electrocuting criminals, there was still much work to be done. Brown's ultimate aim was to have the everyday use of alternating current limited to 300 volts.[54] In January 1889, he circulated, to hundreds of businessmen and government officials across the United States, a letter that once again attacked alternating current.[55] The letter repeated most of his earlier assertions, but in an attempt to show how the lives of the general public could be endangered by alternating current, Brown claimed that it was not uncommon for telephone and telegraph wires to be burned when coming in contact with alternating-current wires. As a consequence, Brown wrote, any person who touched a telephone or call box ran the risk of being electrocuted. He closed by urging recipients of the letter to support legislation limiting the electromotor force of alternating current to 300 volts. This was the first time, at least in print, that Brown referred to alternating current as the executioner's current, a force he deemed so lethal it should only be used when death was the desired result.[56]

Brown relentlessly kept up the assault on Westinghouse and alternating current. In the spring of 1889, he published a pamphlet entitled "The Comparative Danger to Life of the Alternating and Continuous Electrical Currents." In the preface, Brown argued that the "criminal economy" of the alternating current companies had sacrificed many lives and that the state legislature ought to limit the electromotor force of alternating current to 300 volts, except for "the electrical execution of condemned criminals." A 300-volt limit would, of course, effectively eliminate the cost advantage of alternating current. Objecting to charges of "cruelty to animals" levied against him, Brown claimed that his only aim was to save human life, and that the animals killed were already scheduled to die by more painful methods. In order to establish his credentials, Brown thanked Thomas Edison for the use of his laboratory facilities, Edison's chief electrician, Arthur Kennelly, for his expert assistance, and Dr. Frederick Peterson for his indispensable physiological work.[57]

As a way of furthering his case against alternating current, Brown reprinted his controversial *Evening Post* letter. He then complained about the unfair bias and vicious personal attacks he had endured after its publication. According to Brown, criticisms made by the alternating-current people were filled with contradictions and inaccuracies. The most "outrageous" faultfinding came from H. M. Byllesby, vice president of the Westinghouse Electric Company, who asserted that direct current was more dangerous than alternating current.[58] His pamphlet also contained several papers read at the Columbia College School of Mines demonstrations regarding the physiological aspects of the experiments by both Brown and Dr. Frederick Peterson. Brown detailed his experiments with twenty-seven dogs in an appendix.[59] The only new information in the pamphlet was Brown's charge that the Westinghouse Company had prevailed upon Hankinson from the Society for the Prevention of Cruelty to Animals to stop Brown's demonstration before it was completed.[60]

Brown's pamphlet failed to silence his critics; indeed, they grew more numerous and vocal. On July 13, 1889, the *Electrical World* published a devastating critique of Harold Brown by Ludwig Gutmann. The New York electrician argued that Brown's experiments were faulty and misleading. He noted that even though Brown entitled his pamphlet "The Comparative Danger . . . ," he did not perform truly comparative tests.

According to Gutmann, Brown actually compared alternating-current systems without converters to direct-current systems with converters, thus giving DC an unfair advantage. Gutmann also argued that Brown's "automatic safety device" did not make the Edison dynamo safe. In fact, once a live wire was grounded, Brown's device ceased to work and anyone touching it would be electrocuted.[61]

Gutmann went on to assert that Brown had arranged the most favorable conditions for his direct-current experiments, but that none of these conditions is likely to appear in practice. Brown's experiments had been conducted with the express purpose of persuading the public that alternating current was dangerous. At between 200 volts and 1,500 volts, should there be a short circuit, the direct current changes its nature and behaves exactly like alternating current. Although Gutmann acknowledged that the installation of electrical equipment was often done poorly, he pointed out that AC killed very few people compared to home construction or railroad travel. And because of a well-designed converter, AC is much safer than DC. This, Gutmann said, was certainly illustrated in Europe, where there was a higher percentage of skilled workmen and fewer accidents as a result. In America, the introduction of alternating current had been such a commercial success that inexperienced workers were hired in order to meet the demand. Also, most major cities—where the placement of telegraph and telephone wires was unregulated—had a tangle of cables overhead that was at constant risk of coming in contact with high-tension electrical wires. To blame alternating current for this situation, Gutmann said, was patently unfair.[62]

Gutmann acknowledged that Dr. Frederick Peterson, the electrician who assisted Brown at Columbia College, was correct in saying that alternating current would be as dangerous as a direct current of twice the magnitude, but argued that Peterson's conclusion was incorrect. As soon as a ground or short circuit occurred, according to Gutmann, the DC changed into an oscillating current and a 1,000-volt oscillating current would become just as deadly as an alternating current of the same voltage. It was clear, Gutmann insisted, that Brown tried to fix the outcome of his experiments by inserting a deflecting device in the direct-current apparatus to make AC appear more deadly. Brown claimed that he was only interested in the development of engineering and science, but Gutmann charged Brown with impeding progress rather than promoting it. In closing, Gutmann claimed that any fair-minded electrician must conclude that the future belonged to alternating current.[63]

Throughout the public debate and passage of the electrical execution law, the electrical profession had remained curiously silent on the desirability of using electricity to execute criminals. Now, with William Kemmler sentenced to die by electricity, the profession began to speak out. At the August 14, 1889, meeting of the Electric Light Convention, Mr. F. A. Wyman, a Boston attorney, delivered a thunderous condemnation of electrocution. After providing some background information, he declared that nearly every state in America had placed limits on the power to punish criminals for their crimes, and that the Eighth Amendment to the U.S. Constitution prohibited the infliction of cruel and unusual punishment. Wyman deemed death by electricity unconstitutional and predicted that the new law would be repealed. There was little doubt that death by electricity was cruel punishment, Wyman contended, and it was certainly unusual. He noted that the state of knowledge was such that not even the experts could ascertain how much electricity it takes to kill a man. Since men differed in their "powers of resistance," an electrical force that killed one man might not have done much more than injure another, he said. From those who had survived accidental encounters with electricity, Wyman persisted, it was clear that its victims suffered horrible pain. He called for nothing less than a repeal of the death-by-electricity legislation.[64]

After Wyman was done speaking, a lively discussion ensued. Dr. Otto Moses, whose opinion had been solicited by the Gerry Commission, stood to make the observation that while the intent of the new law was to produce a more humane execution, the method by which the execution was to be accomplished was one of the "most barbarous." He complained that the electrocution law had been passed hastily, without adequate input from leading electricians. It was only at the eleventh hour, he noted, that the electricians learned that their science of electricity was to be used as an instrument of death. He noted that at first the electricians had dismissed charges that "rival interests" in the electric lighting industry—"especially one individual"—were behind the campaign to adopt electricity as the mode of execution, but said recent developments in the Kemmler case had convinced him otherwise.[65]

Speaking in a firm voice, Moses quoted from a survey of accident victims that illustrated the painful injuries caused by contact with electrical currents not sufficient to cause death. He said victims reported a wide range of unpleasant sensations after coming into contact with electrical current.

No two people agreed as to the exact symptoms. For instance, it is declared that the action of the current upon one man was like cutting him in two with a buzz saw; another one says it was like striking him on the back of the neck with a sledge hammer; another, like knocking his brains out with a trip-hammer; another said that he felt himself full of needles; another, as if he had fallen from an immense height and had been dashed to pieces upon the ground; another, as if he had been smashed.[66]

Dr. Moses predicted that electrical executions, with their barbarity, would bring disgrace to the electrical profession—a horror worse than the Spanish Inquisition. He said electricity was the only mode of execution that "simultaneously tortured every nerve of the body" and that it was, without a doubt, the cruelest punishment. He maintained that it was crueler than the rack, crueler than being drawn and quartered, and crueler than hanging by one's thumbs, for all these gruesome punishments never "took every nerve in your body and put it upon the stretch at the same instant."[67]

Dr. Moses then appealed directly to the self-interest and pride of those attending the convention. He said the citizens of New York had already developed a fear of electricity, and to execute a man with an electrical current could only serve to heighten those fears. He called on his colleagues to use electricity to "rejuvenate the world." He said electricity was a "civilizing agent, not an instrument of torture." Moses added that electricians should follow the lead of those in the medical profession and refuse to act as executioners. He then said that even as he spoke, there was an electrician at the penitentiary in Auburn preparing for the eventual application of the deadly current to the prisoner Kemmler.[68]

Moses reminded the audience that the dangerous use of electricity for the taking of life was not new. The renowned Benjamin Franklin experimented with turkeys and dogs in order to acquire knowledge for the killing of criminals. During one such experiment Franklin accidentally touched the conductors of a battery; a "most horrible" shock was felt throughout his entire body. Moses ended his speech by offering a resolution that the law requiring the execution of criminals by electricity be repealed. The law was an "unnecessary degradation of the most useful natural agent that science has ever rendered available for the service of man."[69]

In response, Charles R. Huntley, an electrician who would later act as an official witness to Kemmler's execution, rose to introduce Dr. George E. Fell to the convention. The practicing Buffalo physician and surgeon had been assigned to assist Harold Brown in designing the electric chair. Fell claimed that his experiments proved that electricity was both swift and painless, and he remained convinced that a man could be dispatched as easily as a dog. Fell stated that he did not believe the use of electricity for the execution of criminals would hurt the progress of the electrical industry.[70]

The former president of the American Institute of Electrical Engineers, H. W. Pope, was next on the agenda. In a terse reply to Fell, Pope pursued the idea that electrocution would damage the prestige of the electrical industry, stating that the physicians banded together to oppose lethal injection because they felt it would hurt their profession. Pope asked Fell what he thought of the fact that the Society for the Prevention of Cruelty to Animals had recently begun using gas to put unwanted animals to a painless death, and he proposed that the same method might be useful for convicted criminals. Fell responded that the inhalation of poisonous gases was certainly painful, especially in the first few breaths before asphyxiation occurs. Fell maintained that gas was more painful than electricity because it did not travel as quickly through the body.[71]

Wyman, who had delivered the convention's opening address, then announced that it was time to "get down to business." He declared that the public debate on electrical execution could only have a detrimental financial impact on the industry. Members of the National Electric Light Association could no longer allow the present dispute between Edison and Westinghouse to destroy the entire industry. Passing resolutions would not be enough, he claimed, and he urged members to travel to Albany to pressure the governor into commuting Kemmler's death sentence.[72]

A member of the association, Mr. DeCamp, then asked Dr. Fell what evidence he had that an electrical execution could be done "scientifically, without pain." Fell repeated his assertion that the electrical current traveled much more quickly than nerve impulses, which produced a painless death. Professor Anthony followed up on this assumption. He charged that all of the Brown and Edison experiments were unreliable, and it was "not the velocity of electricity [that counts] but simply the time required for the wires to be charged, and for a spark to leap across the gap between the two wires." There was slight applause, and the professor continued. "The fact is that in the nerves of the human body, the velocity of the electrical current is known to be immensely less than it is in a metallic wire," he said. Until the exact speed at which the current traveled through the human body was known, it would be impossible to determine whether or not death by electricity could be painless.[73]

The professor then made the point that an electrical execution required the participation of electricians. He said sheriffs simply lacked the technical knowledge to operate and maintain the elaborate machinery. Professor Anthony speculated that even in one hundred years, electricity would still not be understood well enough by the layman that a sheriff could assume the full responsibility of carrying out an electrical execution.[74]

Finally, convention participant Mr. Morrison urged the members of the association to speak "in good, plain, straight English" in debating the issue at hand. Despite all the rhetoric to the contrary, he claimed, electrical executions were not developed for humanitarian reasons. There was little or no concern for the suffering of the criminal. Without explicitly naming Edison, he claimed that electrocution was created to benefit an individual's private interest and "gratify his personal malice in pitting one system of electrical lighting against another system." Morrison's remarks were met with applause. Encouraged, Morrison claimed that natural gas was the most dangerous source of energy available. Indeed, he noted, most items used in everyday life could cause death or injury if used improperly. In this sense, there was nothing unique about electricity. He claimed electricity could not produce a painless death because the dreadful appearance of the chair the condemned man would sit in was almost enough to scare him to death. Morrison closed his remarks by urging members of the National Electric Light Association to do everything in their power to repeal the new electrical execution law.[75]

The meeting ended with a resolution to appoint a committee to present a repeal petition to the New York state legislature.[76]

ANY HOPE that Harold Brown had of maintaining his image as a public-spirited independent investigator was completely destroyed by the publication of his private letters in the *New York Sun*. Someone had burglarized his office at 45 Wall Street and stolen nearly fifty letters linking him both financially and strategically with Thomas Edison and Charles A. Coffin of the Thomson-Houston Electric Company.[77] The *Sun* gave these private letters a prominent place in its Sunday edition on August 25, 1889. In bold letters, its headlines read,

FOR SHAME, BROWN!
Disgraceful Facts About the Electric Killing Scheme.
QUEER WORK FOR A STATE'S EXPERT
Paid by One Electric Company to Injure Another.

The accompanying editorial identified Brown as the driving force behind the state of New York's adoption of alternating current for use in electrical executions. The *Sun* predicted that, like Joseph-Ignace Guillotin in France, the name of Harold Brown would be forever associated with the electrical method of execution. Since the beginning, the *Sun* stated, Brown had pretended that his only interest in electrical execution was as a scientific investigator, a "humane man" merely seeking the "quickest and most painless method" of putting condemned criminals to death. Although his popular image was that of a "public-spirited individual with a fondness for dabbling in electricity," the *Sun* charged that Brown's real motive was always to destroy the alternating-current industry. The editorial further noted that, as a man of moderate means, even his friends had wondered how Brown could have devoted his full time to the campaign to sabotage the alternating-current industry.[78]

The letters themselves revealed that, from the beginning, Brown had the financial backing and technical support of Thomas A. Edison. The Edison Company publicized Brown's mischievous challenge of Westinghouse to an electrical duel, along with an early version of his notorious "Comparative Danger" pamphlet, which they sent to state lawmakers in Missouri and elsewhere. The letters also confirmed that Edison and Brown were conspiring together in Pennsylvania to limit the transmission of electricity over high-tension wires to 300 volts, and that Brown's animal experiments were conducted under expert staff supervision at Edison's West Orange laboratory in New Jersey.[79]

Brown's letters further documented the elaborate scheme and subterfuge that he and Edison engaged in to purchase a Westinghouse dynamo for use in the first electrocution. Although the Medico-Legal Society had recommended alternating current as the executioner's current, it did not specify which manufacturer's dynamo it favored. The law left the purchase of the necessary electrical apparatus to the superintendent of state prisons. A letter indicated that Harold Brown was asked to present an "itemized estimate of the cost" of an execution apparatus at a meeting on March 20, 1889, to Superintendent Austin Lathrop and the wardens of Sing Sing, Auburn, and Dannemora state prisons, as well as a bill for his recent experiments at the Edison laboratory.[80]

At that meeting prison officials authorized Brown to secure and install a fully functioning electric chair at all three locations, but they refused to pay Brown "until the first execution proves that plant is suitable for the [intended] purposes." Lacking funds of his own, Brown turned to Edison for the money. He informed the beleaguered inventor that an "investment" of $7,000 to $8,000 would be required and he fretted about having to pay for "at least one full complement of converters and lamps," presumably because the simple purchase of a dynamo would arouse undue suspicion.[81]

Brown also indicated to Edison that the Thomson-Houston people — who were negotiating a merger with Edison Electric Light Company at the time — had authorized him to take up Westinghouse's challenge that his alternate-current lighting system was 50 percent more efficient than any direct-current system.[82] Brown promised to undertake both projects if $5,000 was made available to him and explained to Edison that his prior approval was necessary before the Thomson-Houston Company would pay him. "By showing the lack of efficiency of the Westinghouse apparatus, [we can] head off investors and prick the bubble, thus helping all legitimate electrical enterprises."[83]

The Westinghouse Company and its various affiliates were well aware of Brown's intentions and were not about to sell him any of its dynamos. With the help of Charles A. Coffin, the treasurer of the Thomson-Houston Electric Light Company, however, Brown managed to purchase three Westinghouse dynamos for use with New York State's electric chairs. The documentation is sketchy, but Brown's letters indicate that Coffin recruited Frank Ridlon, a secondhand-electrical-apparatus dealer in Boston. Ridlon was president of Frank Ridlon & Company as well as manager of the Baxter Manufacturing and Motor Company, both located at 196 Summer Street in Boston. It was Ridlon who actually obtained the Westinghouse dynamos, probably through the Oneonta Electric Light and Power Company of New York. In any event, with Edison's approval, Coffin authorized payment to Brown of $1,200 for each of the dynamos,[84] $1,000 for Brown's efforts at Johns Hopkins University to demonstrate the inferiority of Westinghouse dynamos,[85] and an additional $500 for "future services."[86]

During the complex negotiations Brown often expressed concern that the Westinghouse people would expose his plan and defeat his strenuous efforts. He promised Coffin that the Thomson-Houston Company's name would be kept out of official documents; he sought assurances from Frank Ridlon that the dynamos would be shipped in unmarked crates, and from

Warden Charles Durston he sought assurance that the dynamos would arrive at Auburn without attracting notice. Brown was especially concerned about rumors that the Westinghouse people would try to sabotage the dynamo once it was installed at Auburn. He wrote Coffin warning of a "desperate attempt" to stop the use of the Westinghouse dynamo. He said that he knew of the Westinghouse Company "offering money to some of the prison officials . . . and that they have someone at Auburn."[87]

As might be expected, the damning evidence disclosed in the *Sun* letters unleashed a flood of public condemnation. The following week an editorial in the *Electrical World* harshly criticized Brown. After the *Sun* published Brown's confidential letters, the *World* declared, it was clear that his interest in electrical executions, despite his insistent claims to the contrary, was neither public-spirited nor philanthropic. Brown's true aim, said the *World*, had always been to use a Westinghouse alternating-current dynamo for electrical execution in order to protect his interest in direct-current systems. Indeed, through one contrivance or another, the editorial noted, Brown had made quite a bit of money for himself. According to the *World*, Brown was the man who carried the campaign to discredit alternating current, claiming that execution by alternating current would be "marvelously quick and painless." Finally, they said, the confidential letters showed that Brown was telling the public one thing and the electric light companies who supported him quite another.[88]

The *Electrical World* was hardly unique in its criticism of Brown's secret corporate sponsorship. What was interesting about both the *World* and the *Sun* editorials was their failure to place the blame where it actually belonged—on the shoulders of Thomas A. Edison. Even though Brown was exposed as a front man for the Edison Company, Edison himself managed, somehow, to escape any real public condemnation.

Amid the growing criticism of Brown, Eldridge Gerry felt the need to publish a defense of the new method of electrical execution. In a September 1889 article in *North American Review*, Gerry reminded readers that America had always utilized the death penalty in one form or another, and until recently the production of pain was considered an important component of punishment by death. Under the Mosaic code, he noted, people acted as executioners by stoning offenders to death. As civilization advanced, however, the taking of life without the infliction of unnecessary pain or suffering came to be regarded as the appropriate philosophy of execution. This had not reduced the deterrent effect of capital punishment, Gerry contended, for the criminal feared death itself more than the "mode of its infliction."[89]

In explaining his commission's report, Gerry said that America inherited hanging from its mother country, but a careful consideration by New York legislators in 1886 deemed hanging "uncertain and dependent partly on the nerve of the convict and partly on the skill of the executioner." Gerry said that if the neck was not broken with a sudden jerk, the condemned strangled to death, a process that could take up to half an hour.[90]

Gerry then recounted cases in which the rope broke, decapitations from the use of too small a noose cord, and "authentic cases of subsequent resuscitation." According to Gerry, all of these demonstrated an urgent need to find a practical, quick, and painless method of execution, one that was "uniform in its results." He noted that of all the forces known by science to produce death, electricity was the quickest. He maintained that hanging, while not unusual, was probably cruel, and he argued that electricity, while not yet usual, produced death so quickly that it was difficult to imagine how it could be defined as cruel.[91]

Brown, though shaken by the publication of his private letters, remained steadfast. Two issues later, in November 1889, he published an article in the *North American Review* entitled "The New Instrument of Execution."[92] Brown's article was both a passionate defense of his killing experiments with animals and a sustained argument for their application to the execution of criminals. After complaining that he was "fiercely and bitterly attacked" by proponents of alternating current, he expressed his sympathy for Thomas Edison, whose enemies had not forgiven him for making his "magnificent" laboratory available for scientific experimentation. Against all evidence, Brown claimed somewhat disingenuously that he was at first reluctant to accede to the request by New York State authorities to procure the electrical apparatus for execution, knowing that it would expose him to verbal abuse. It was, however, an opportunity to save many lives by demonstrating that alternating current was really harmful and unsafe.[93]

On the crucial question of the selection of a Westinghouse dynamo, Brown asserted that he decided against designing a new "machine" because there was already available a dynamo with proven ability to destroy life. Several electricians and physicians had affirmed publicly that the resistance of the human body could not be measured accurately, but, according to Brown, this was not true. He noted that at the West Orange, New Jersey, laboratory, Edison had measured the resistance of four hundred to five hundred men; their average resistance was 1,000 ohms. And several experiments by Arthur Kennelly proved that the current itself reduced the resistance of the skin by attracting the fluids of the body.

Therefore, a quick and painless death merely depended upon adequate contact points. Brown pointed out that electricity killed primarily by causing violent vibrations in the body fluids, which destroyed the vital organs. A major characteristic of death by electricity was the darkening of the blood, which appeared to be caused by loss of oxygen; Brown said scientists did not know whether this change preceded or followed death. Nonetheless, Brown maintained that death by electricity was quick and painless, and any pain or suffering could only be the result of imperfect contact. He wrote that because electricity traveled at the speed of light and nerve impulses only travel at 180 feet per second, the criminal was dead before the sensation of pain could reach his brain.[94]

Aware that critics had objected to preparations necessary for an electrical execution as too complicated, Brown stated that they are in fact very simple.

> The condemned criminal's cell is visited by the prison authorities and his hands and feet are saturated with the weak potash solution which so rapidly overcomes the skin's resistance; during this space of thirty seconds or less his electrical resistance may be measured, though Mr. Edison's researches in this line have rendered even this unnecessary. Shod in wet felt slippers, the convict walks to the chair and is instantly strapped into position; his feet and hands again immersed in the potash solution contained in a foot-tub connected to one pole and in hand-basins connected to the other. With this perfect contact there is no possibility of burning the flesh and thus reducing the effect of the current upon the body.[95]

Brown ended by chiding the alternating-current companies for attempting to prevent the legislature from outlawing its current and thus "ending the terrible, needless slaughter of unoffending men."[96]

For the same issue of the Review, Edison was asked to write a short article about how to avoid the dangers of electric wires in the street.[97] Instead, Edison seized the opportunity to launch another attack on alternating current and its purveyors, even though the Thomson-Houston Company had just entered into an agreement to purchase the entire plant, business, and patent rights of the Brush Electric Light Company, an alternating-current enterprise.[98] In an article entitled "The Dangers of Electrical Lighting,"[99] Edison told of an accidental electrocution in New York City. He warned that as electric lighting became more widespread and the number of wires in a city multiplied, more accidental deaths could be expected. According to Edison, high-tension currents had killed nearly one hundred people, and the only reason accidents were more plentiful in New York City than anywhere else was that New York had more wires per square mile than any other city in America. Edison confidently predicted that as other cities expanded their usage of electricity, their accident rates would begin to rise.[100]

Many politicians and electricians had suggested that putting high-tension wires underground would make them safer, but Edison said he believed this temporary fix would actually make the system more dangerous. He argued that even a perfectly insulated wire carrying a high-tension current would not guarantee safety because eventually vibrations would cause defects in the wire's insulation. The only real solution, as Edison saw it, was for the political authorities to regulate electrical pressure—that is, reduce it to harmless levels below 300 volts.[101]

In every way, Edison claimed, direct current was superior to alternating current. He alleged that companies used alternating current merely to reduce investment in copper wire and real estate. The weight of copper wire necessary for four circuits of ten lamps, he noted, was two and a half times greater than what was necessary for one circuit of forty lamps. Therefore, companies used AC because they did not want to make the investment in copper that DC required; also, the real estate cost of constructing a central power station was less expensive for alternating-current manufacturers. Stations producing low-tension current had to be located in the center of the city near their customers; alternating current used a high-tension transmission of current that easily traveled longer distances, so its plants could be located on the outskirts of the city where land was cheap. In closing, Edison pointed out that in New York City, a harmless direct current of no more than 220 volts served thousands of customers. Therefore, he argued, since central stations could operate at a profit, there was no excuse for endangering the lives of New Yorkers by manufacturing the dangerous alternating current.[102]

George Westinghouse mostly stayed out of the public debate on the relative merits of alternating and direct current, but by the autumn of 1889 he had heard enough. Finally, in November 1889, the Westinghouse Company wrote to the *Electrical Engineer*, complaining about the uncritical publicity they had been providing Harold Brown. The letter caused the journal to investigate Brown's allegations that the Westinghouse system of alternating current was responsible for thirty fatalities during the previous year. Brown accused Westinghouse by name in fifteen cases; in the remaining fifteen no particular company was named. The *Electrical Engineer* reported that their own investigation had uncovered the following facts: of the thirty cases, one, possibly two, could be attributed to the Westinghouse system. In twelve of the remaining twenty-eight cases, the journal reported, there were no Westinghouse plants in the area at the time of the accident. And in sixteen cases, the reports were either untrue or due to alternating-current systems run by other companies. The *Electrical Engineer* concluded by encouraging all other alternating-current manufacturers to join them in challenging Brown's outrageous assertions.[103]

AVARICE, CORRUPTION, AND OUTRIGHT DISHONESTY effectively shaped early debate over the electric chair. The lawmakers who ultimately decided on the chair were informed by faulty scientific experiments conducted by self-interested inventors and by electrical and medical experts who were largely ignorant of the effects of electricity on the human body.

This story does not end with the passage of the electrical execution law or the invention of the electric chair. Before the chair could claim its first victim, a number of formidable legal hurdles had to be cleared, the result of which resulted in Kemmler's case establishing legal precedent on the constitutionality of electrocution.[104] Although his contributions to legal history were immense, the circumstances of William Kemmler's life were much less significant than those of his death. Nonetheless, his story is worth telling, for it provides a window into the social, cultural, and political environment that gave rise to the electric chair.

CHAPTER FIVE

"I'll Take the Rope"

The Life, Crime, and Trial
of William Kemmler

WILLIAM FRANCIS KEMMLER was born on May 9, 1860, to German Lutheran parents who lived on the outskirts of Philadelphia at 2531 North Second Street. He was one of eleven children, of whom only two survived to adulthood. Although Kemmler attended school until the age of ten, he never learned to read or write. As a young boy, William spent most of his childhood in the streets of Philadelphia working to contribute to the income of the family; he sold newspapers, polished boots, and worked in his father's butcher shop. At seventeen, Kemmler got a job at a brickyard, where in just two years he saved enough money to buy a horse and cart. He established a small business peddling fruit and vegetables. It was during this period that Kemmler first demonstrated a "fondness for alcohol," and as his business prospered, he spent his money freely. He was often seen in the company of men with a similar thirst for whiskey.[1]

One night while he was in a drunken stupor, a young woman named Ida Forter seduced Kemmler and persuaded him to marry her the following day in Camden, New Jersey. During their honeymoon, Kemmler discovered that Forter had been previously married but never divorced. Two days after the ceremony, bitter and uncertain whether his marriage to Forter was legal, Kemmler left his young bride and took up with a married woman, Matilda Ziegler. Tillie, as she was called, was the runaway wife of Fred Ziegler, a railroad man and cabinetmaker in Philadelphia.[2]

(*New York Herald*, 7 August 1890)

Matilda "Tillie" Ziegler

Tillie and Fred Ziegler got along well at first. But after their daughter, Ella, was born, the couple began to experience trouble in their relationship. After nine years of marriage, Tillie left Fred because of his "fondness for the society of lewd women," his "quick-temper," and his failing health. Tillie was willing to tolerate Ziegler's behavior as long as he worked and provided for her and Ella. However, when he became ill with "severe hemorrhages" and a "loathsome disease," she decided to leave him. Alone and penniless, Tillie sought comfort and support in the arms of William Kemmler.[3]

Less than ten days after taking up with Tillie, Kemmler sold his business and property for $1,200, and the couple moved to a German section of Buffalo, where they rented the back portion of a house at 526 South Division Street. Their home in this large, single-story, old-fashioned square cottage consisted of four dingy rooms in the rear of the building. Kemmler was attracted to the dwelling because it contained a small detached barn where he could keep horses and chickens. Once settled in his new home, Kemmler registered for his peddler's license under the name of John Hort, a man he had known in Philadelphia. Until Tillie's murder, the coupled lived as man and wife under the names John and Matilda Hort.[4] Kemmler would later tell police that because he and Tillie were unmarried, he took out his license in the name of John Hort. He probably used the alias because he feared that if Tillie's husband regained his health, he might come looking for them.

All was not well between Kemmler and his paramour. During the eighteen months that Kemmler and Ziegler lived together in Buffalo, they quarreled frequently. On more than one occasion, Kemmler discovered that Ziegler had entertained male friends at their home during his absence. The couple argued repeatedly about Kemmler's drinking binges, Ziegler's infidelity, and Kemmler's claim that Ziegler often stole money from him. In the weeks leading up to Ziegler's murder, Kemmler suspected her of having an affair with his partner, John De Bella, who lodged with them.[5]

In the early morning of March 29, 1889, Kemmler was about the barn directing his employees and attending to his horses. He was sober, although he had been up late drinking the night before. As she often did, Ziegler emerged from their flat and asked for some fresh eggs. When she returned to the house, Kemmler followed her. While she was washing the breakfast dishes, he attacked her with a hatchet, delivering twenty-six blows to her head, neck, and chest. Little Ella, paralyzed with fear, watched the entire thing.[6]

His inner rage satisfied, Kemmler rushed into Mrs. Mary Reid's rooms at the front of the house, his sleeves and hands covered with blood. "I have killed her," he exclaimed loudly. "I had to do it; there was no help for it. Either one of us had to die. I'll take the rope for it." In disbelief, Mrs. Reid asked Kemmler if he really meant what he said. "Yes I do; see my hands," he replied. They were dripping with blood. A few moments later, Ella rushed into the room and cried out hysterically, "Papa has killed Mamma." Mrs. Reid ran for help. Meanwhile, Kemmler took his four-year-old stepdaughter by the shoulders and calmly placed her in a chair. He washed his hands and left the cottage.[7]

Kemmler headed toward the street, but was met at the front gate by Asa King, the adult son of a neighbor. King pleaded with Kemmler to go with him to find a doctor, but Kemmler refused, saying, "I have killed her, and I expect to hang for it." Kemmler then headed for the corner bar-room, Thomas Martin's Saloon. He called for a drink. King, who had followed Kemmler, told the bartender not to give him any liquor. While the two men quarreled, Officer John O'Neil of the Third Precinct appeared, and having been notified of Kemmler's alleged crime, promptly began questioning the suspect.[8]

Officer O'Neil asked Kemmler if he had been "licking his wife." When Kemmler answered in the affirmative, the officer asked him with what. "With a hatchet," was Kemmler's reply. Upon the officer's request, Kemmler showed O'Neil the way to his house. Tillie Ziegler's mutilated body was lying on the kitchen floor in a pool of blood. The murder weapon, a bloodstained hatchet, lay nearby. The kitchen was littered with broken dishes; tables and chairs were in total disarray. To Officer O'Neil's surprise, Ziegler was still alive—her arms and legs were flailing back and forth. She was taken to nearby Fitch Accident Hospital, where she died at 12:50 the following morning.[9] Ziegler's skull was fractured in five places, and her upper torso was covered with deep gashes. A handsome, buxom brunette in life, Tillie was horribly disfigured in death. She was just shy of thirty-one years old.[10]

At the station house, Kemmler remained cool and composed. He gave his name as William Kemmler, but refused to provide the police with a motive for his crime. He expressed no remorse, and he attributed the murder to his "quick temper." He told Lieutenant Daniel E. Barry, the officer in charge of the police interrogation, that he had been drinking heavily during the preceding winter. When asked how he had attacked Ziegler, Kemmler replied, "With a hatchet. I wanted to kill her, and the sooner I hang, the better it will be."[11]

ON MAY 6, 1889, jury selection was conducted in the Court of Oyer and Terminer of Eric County. The Honorable Henry A. Childs presided. During jury selection, Kemmler sat in back of his counsel. He wore "an appearance of dejection . . . an excellent imitation of a man about to drop dead." He had on the same brown suit he had worn during his five weeks of confinement. His face exhibited "the brown hue of the children of the forest." Since his arrest, his face had "paled somewhat, but the "banners on the side of his head still gleam like the side of a coupe." He continued to display "the same appearance of mild-eyed imbecility." Kemmler's brother, Henry, sat next to him, but his sister-in-law kept her distance.[12]

By two o'clock the courtroom was filled with potential jurymen whose names had been drawn from a list of registered voters. There was also the usual crowd of curious onlookers who scurried "to a murder trial as to a picnic." At the outset, numerous potential jurymen were released because "crops would not permit them to leave home for such a trivial matter as a murder trial."[13] Others were excused for reading local papers, such as the *Amherst Bee*, and because they were related to court and jail officials. By the early evening, however, twelve men had been selected to be Kemmler's "life and death arbiters." Among them were men with "tender consciences from South Wales, citizens burdened with scruples against capital punishment from Amherst," agriculturists or farmers who were Quakers, and "a cheese-maker from North Collins who didn't know the difference between capital punishment and a knock-out with four ounce gloves."[14] Today, anyone with scruples against capital punishment for personal or religious reasons would not be permitted to sit on a jury in capital cases.[15]

Most of the jurymen were selected because they answered the court's questions satisfactorily.[16] However, Hiram Emerson, a carpenter from Concord, created a stir in the courtroom when he told the court that he "didn't believe the present law was just what it ought to be, but was unable to see that the defect was his fault." Nonetheless, Emerson was selected because "he thought he could give a verdict in accordance with the facts." During it all, Kemmler appeared largely uninterested in the questioning of potential jurors. For the entire afternoon he sat motionless with his "eyes cast down and his body bending forward until his head was almost level with the table." He only moved when ordered to face a juror about to be sworn. Then, he stared piercingly at each of them, creating a general feeling of discomfort in the courtroom. Intimidated, most of the jurymen avoided Kemmler's eyes, looking instead out the window or at the ceiling.[17]

The entire process of jury selection lasted only four hours. By six o'clock Judge Childs had excused the panel. The jurors were given the usual cautions about discussing the trial with friends and relatives. They were not sequestered.

On May 7, 1889, the actual trial began. At 9:40 a.m., the twenty-eight-year-old defendant entered the courtroom, handcuffed. He had dark brown hair and a light brown mustache. Kemmler was slender, five feet ten inches tall, and about 160 pounds, with slightly stooping shoulders.[18] He took his seat beside his counselors, Charles S. Hatch and Charles Sickmon. Bright sunlight streamed in through the open shutters of the Buffalo courtroom while Kemmler sat with his head down, elbows resting on his knees, "the most indifferent picture of dejection."[19]

A young woman looking very much like "the victim of that morning's work" sat inside the bar of the courtroom. She was a "small, plump [woman], neatly dressed in black," about thirty years old, and extremely

(New York World, 8 August 1890)

William Kemmler

attractive. Her name was Bertha Tripner; she was both the sister of the murder victim and the sister-in-law of the defendant. Bertha occupied what the *Buffalo Evening News* called a "peculiar position. She must have a natural feeling of revenge against her sister's betrayer and slayer; she may, perhaps, sympathize with the brother of her husband who stands in such deadly peril." Kemmler's brother, Henry, sat attentively beside her.[20]

DISTRICT ATTORNEY GEORGE T. QUINBY opened the case for the state. He began with a vivid description of Ziegler's brutal murder with a hatchet. Since there was little doubt that Kemmler committed the crime, Quinby concentrated his efforts on obtaining a conviction for murder in the first degree. He told the jurors that the evidence would show "deliberation and premeditation." And to fend off any feeling of sympathy the jurors might have for the defendant, he emphasized Kemmler's stormy, emphatic statement immediately after the murder, "Yes, I have done it, and I am willing to take the rope for it."[21]

(*Buffalo Commercial Advertiser*, 10 May 1889)

District Attorney George T. Quinby

Next, Quinby stated that Kemmler's jealousy of his partner, John De Bella, provided motive for the murder. Playing on the jurors' fear of crime he warned, "There is getting to be a frightful number of homicides and the punishment meted out to the murderers does not seem to check the crime. It is time that such a salutary lesson should be taught as will have a deterring effect." Quinby then attempted to undercut the defense by predicting that the defense would be driven to the strategy of last resort: "Temporary insanity," Quinby exclaimed, "They will ask you to believe that this man was sane one second and insane the next."[22]

The prosecution first called William J. White, a civil engineer whom the District Attorney's Office hired to make a sketch of the murder scene. He testified that the diagram on display in the courtroom was correct.[23] More important, he testified regarding the conditions of the South Division Street apartment at four o'clock on the afternoon of the murder. Pointing to the diagram of Kemmler's apartment, White began his testimony by saying that the "three great red splashes in the kitchen stood for the blood of the victim." Mr. White also noted five bloodstains on the papered wall. He said that "they were caused by bloody hands being pressed against it" in an apparent attempt to escape or to get help. According to White, "there were also blood stains upon many dishes on the floor," an overturned table, and other signs of a struggle. "The tray of a trunk, filled with articles of women's wear, was also upon the floor," suggesting that Tillie Ziegler was planning to leave Kemmler.[24]

After Mr. White left the witness stand, postmortem examiner Dr. Henry Bingham, the coroner, John Kenney, and Dr. Roswell Park, consulting physician at Fitch Accident Hospital, all failed to answer when called to testify. The next witness was Dr. Charles S. Jones, house surgeon at the Fitch Hospital. He testified that he first saw Ziegler when she was brought into the hospital at 9:00 the morning of the murder. She was still alive. Ziegler was "covered with blood to quite an extent . . . and there were wounds upon the head. She was apparently unconscious." Jones said he found twenty-six wounds in all. He also testified that Dr. Park extracted seventeen pieces of bone from her fractured skull. Upon Quinby's direction to do so, and over the defense's objection, the witness showed the pieces to the jury. The defendant Kemmler sat in silence. His head now resting on the table before him, Kemmler appeared uninterested in the extent of Tillie's injuries. On cross-examination, Charles S. Sickmon, one of Kemmler's attorneys, oddly asked Jones if Ziegler had been an attractive woman. "Average-looking," he replied.[25]

By this time, Dr. Roswell Park had arrived in the courtroom. He had attended to Ziegler while she was in the hospital. He examined her between 11:00 a.m. and 12:00 p.m. on the day of the attack, while she was unconscious. He testified that in a desperate attempt to save her life by relieving the pressure on her brain, he removed bone fragments from her skull. He also said that the wounds were made by a blunt object, and when shown the alleged murder weapon by Quinby, Park testified that the back part or the hammer part of the hatchet was a blunt object. He told the court that he regarded Ziegler's case as "hopeless" and that any wound inflicted by that instrument would be "dangerous to human life." This established that the intention of the attacker was to cause the death of the victim, a necessary element for a first- or second-degree murder conviction. Perhaps trying to establish that Kemmler had good reason to be jealous, on cross-examination, Sickmon once again asked if the victim was attractive. "Not as I saw her," replied Park.[26]

After questioning John Kenney, the coroner, and Thomas Corlett, a medical student who arrived at the murder scene with the ambulance, the prosecution called Asa D. King, "a gray-haired, gray-whiskered book binder from Boston."[27] At the time of the murder he was visiting his father, Earl, who lived next door to Kemmler at 530 South Division Street. According to King, Mrs. Reid ran into his house with her two children, "giving the alarm." King testified that he went to Kemmler's house, opened the kitchen door, and found Ziegler. She was "on her hands and knees, or knees and arms," her bloody body "swaying backwards and forwards, and her long hair hanging down." King said he shut the door and opened it again to find Kemmler standing there wiping blood from his hands. At this point, District Attorney Quinby suddenly shouted, "Kemmler, look up." Kemmler slowly raised his head from the table and with a dazed look gazed vacantly at the district attorney. "Is that the man?" "That is the man," King replied as Kemmler's head sank down again.[28]

After things calmed down, King testified, "I said to Kemmler, 'This is brutal, man. Let's go for a doctor.'" But, instead of getting a doctor, Kemmler stepped over Tillie's body and headed out the door toward Martin's Saloon for a drink. King followed him and told the bartender not to serve him. "He has brained his wife and she is now weltering in her blood." According to King, Kemmler turned to him and said, "Yes I have done it, and I am ready to take the rope for it; the sooner the better."[29]

Unable to get served, Kemmler left the saloon in search of a place to drink. Once again, Mr. King followed him, shouting, "Does the doctor live down here?" Kemmler turned to King and said, "Let me get a glass of beer and I will go for the doctor." After coming out of the saloon, the two men headed toward a drugstore. Again, King asked Kemmler if he would go for a doctor. This time Kemmler said flat out, "No, I won't," and took off in search of another saloon.[30]

Kemmler's confession now in evidence, Quinby attempted to tie Kemmler to the murder weapon. He asked King if Kemmler was holding anything when he saw him wiping his hands. Much to his surprise, King answered, "No, sir."[31] Quinby had broken the cardinal rule of witness examination—never ask a question that you do not know the answer to—but the evidence was so strong against Kemmler that it would not make any difference.

On cross-examination, Sickmon tried to poke holes in King's testimony but to no avail. All he succeeded in doing was underlining the brutal nature of Kemmler's crime and his indifferent attitude toward the suffering of the woman he supposedly loved. Despite Sickmon's best efforts, King managed to portray Kemmler as more concerned with getting himself a drink than in finding Tillie a doctor.[32]

The court then recessed, and when the trial recommenced at 2:00 p.m., Mary Reid was called to testify. Mrs. Reid said that Kemmler lived in the back portion of her house. Most of her testimony merely supported the testimony of others regarding Kemmler's drinking habits and the couple's frequent quarreling. She told the court that Kemmler confessed his crime to her. When she saw Kemmler come out of his apartment, she said, "Mr. Hort, what is the matter?" He replied, "I've killed her." When Mrs. Reid answered, "You don't mean it," Kemmler said, "Yes, Mrs. Reid, I have and I'll take the rope for it."[33]

Upon cross-examination, Mrs. Reid testified that Kemmler never raised his voice to Ziegler, that it was she who did all the shouting. Ziegler always threatened Kemmler with "You will have to mind me, or I'll leave you." Mrs. Reid further testified that Kemmler, nonetheless, seemed to think a great deal of Ziegler. She used "unladylike language," said Mrs. Reid. "I never heard him speak cross to her."[34]

After Dr. Henry Bingham, the physician who had performed the autopsy on Ziegler's body, testified as to the cause of death, a next-door neighbor, Mrs. Nellie Roberts, testified that she had seen Kemmler coming out of the house "wringing blood from his hands." A little later she saw him on the rear veranda holding his daughter, Ella, whom he set down at Mrs. Reid's door.[35] Patrolman John O'Neil, a seventeen-year veteran of the Buffalo police force, was then sworn in. He told the court that Kemmler appeared sober and did not resist arrest. O'Neil had nothing more to add, except that Kemmler had $365 in silver and copper coins in his pockets when he was arrested, and another $300 or $400 in a trunk at home.[36]

Dr. Delevan E. Blackman, who lived only a short walk from the couple's home, was the first physician to attend to Tillie Ziegler. When he arrived at the scene he saw her lying unconscious on the kitchen floor in a pool of blood. After examining her head wounds, Dr. Blackman picked her up and placed her on a nearby bed. Tillie momentarily regained consciousness and her arms began to flail, splattering blood all over the bedroom wall.[37]

Following a detailed description of the murder scene, Dr. Blackman identified the hatchet in evidence as the same one he saw lying on the floor of the kitchen at the murder scene. When questioned about how he could be sure it was the same short-handled ax, Blackman replied that he had marked it with a notch just below the blade.[38] The bloodstained hatchet was then handed to the jurors for their inspection. Kemmler, who had been sitting quietly staring at the floor, briefly raised his head and cast a quick glance from beneath his brows. Then he let his head drop again.[39]

After two more witnesses testified to matters unimportant, District Attorney Quinby called one of his key witnesses to the stand, John "Yella" De Bella. De Bella was a Spaniard with black curly hair. He was nicknamed Yella because of his jaundicelike complexion and the convenient fact that Yella rhymed with Bella.[40] De Bella told the court that he began boarding with the Kemmlers shortly after he had gone to work for "Billy" the previous January. He said that Kemmler was known locally as Philadelphia Billy, but that his peddler's license was taken out in the name of John Hart. When Quinby tried to correct him, De Bella insisted that Kemmler always spelled his name with an *a*, not an *o*. The matter inconsequential, Quinby went on to establish the nature of their business relationship. De Bella testified that the two men were partners, and although Kemmler owned all the horses and wagons and kept the books, they shared profits equally. De Bella's job was to help with the selling and attend to the care and feeding of the horses. Together they ran two wagons; a third wagon was run by an employee, Charles Spang.[41]

According to De Bella, Kemmler was sober the morning of Tillie's murder, and he had noticed nothing peculiar about his partner's appearance. While counting eggs and transferring them from one barrel to another, De Bella said he had heard something fall in the Kemmlers' kitchen, but did not think much about it. After finishing his work and brushing his clothes, De Bella headed toward the house for breakfast. Once inside, he found Tillie lying wounded on the floor. He immediately ran out to fetch Mrs. Reid. When Mrs. Reid did not answer her back door, De Bella sent Dennis McMahon, a sixteen-year-old boy who worked for Spang, around to the front door to ring her bell. By the time De Bella reentered the apartment, Dr. Blackman was kneeling beside Ziegler's nearly lifeless body.[42]

In an attempt to help establish a motive for the crime, Quinby asked De Bella if he had ever heard Ziegler threaten to leave Kemmler and return to Philadelphia. De Bella admitted that about three or four days before the murder, he had heard Ziegler make such a threat. This remark caused a bit of a stir among the spectators. Kemmler's attorney jumped to his feet to object to this line of questioning. He argued that it was incompetent and immaterial. Judge Childs, however, quickly overruled him, instructing him to sit down.[43]

Next, De Bella testified that Ziegler had said "she was not going to live with a . . . drunken man," that she was going back to Philadelphia. When Quinby asked De Bella if Kemmler had ever said anything about Tillie wanting to return to Philadelphia, Sickmon objected, claiming it was all hearsay. But Judge Childs allowed it. De Bella said that Kemmler told him something along the lines of, "She wants me to go to Philadelphia and when I want to go she don't want to go." Apparently unable to grasp that Ziegler wanted to return to Philadelphia to get away from him, Kemmler told De Bella that he did not understand women, that "women a'hell."[44] After hearing this, Judge Childs told the jurors to disregard the statement.

Quinby then asked De Bella if Kemmler had ever used the expression "putting him in a hole." Once again Sickmon objected. Nonetheless, Judge Child said that he would allow it, at least until he heard some more. De Bella further testified that Kemmler had said this to him on several occasions, but that every time he asked him what he meant by it, Kemmler would reply cryptically, "You know as well as I know."[45]

Sickmon asked that this remark be stricken from the record. In response, Judge Childs asked Quinby to explain how it was relevant. Quinby replied that he was trying to show that Kemmler was jealous of De Bella, that he thought that there might be something romantic between him and Tillie. The expression "putting him in a hole" pertains to this. "It goes to the question of motive." Sickmon again objected, arguing that this particular expression referred only to Kemmler's and De Bella's business partnership. Judge Childs agreed that the expression was ambiguous and could be interpreted in several different ways. Hence, he ordered it stricken from the record.[46]

The district attorney, of course, did not have to provide a motive for Kemmler's murderous attack; all he needed to prove was mens rea, criminal intent. There was no question Kemmler acted with malice aforethought. To ensure a first-degree-murder conviction, however, it is often essential for the prosecutor to furnish the jury with a motive. Jurors need to understand why the defendant committed the crime. Also, with the defense planning to argue alcoholic insanity, evidence of motive was particularly important to the prosecution. Otherwise, in the absence of a motive, Kemmler's crime might seem senseless, the act itself evidence of his diminished mental capacity. Without the sufficient mental capacity to premeditate a crime, Kemmler could not be convicted of murder in the first degree.[47]

It was now Sickmon's turn to attack the state's witness. Upon cross-examination, the court learned that De Bella had lived in Buffalo for less than two years. Previously, he lived in New Haven, Connecticut, where he was in the fruit and vegetable business. Aiming to make his case that Kemmler suffered from alcoholic insanity at the time he committed his crime, Sickmon persuaded De Bella to detail the events of the afternoon prior to Tillie's murder. De Bella acknowledged that he and Kemmler spent the afternoon drinking, visiting one saloon after another until they were dead drunk. When Sickmon suggested to De Bella that Kemmler had "staggered" and "talked at random," implying that Kemmler's conversation was incoherent, Quinby strenuously objected. But Judge Childs said that the witness could go ahead and describe Kemmler's condition.[48]

De Bella's description of Kemmler's behavior seemed to help the defense. He testified that "when he [Kemmler] would take a drink of whiskey in the morning he would follow it up all day, and when he would take beer in the morning he would follow that up all day." Moreover, De Bella said that Kemmler was never sober for more than a week, that "every other saloon he would come to" he would stop for a drink. When Sickmon asked if this meant that Kemmler was intoxicated "more or less every day," De Bella answered, "Oh, some days he would keep sober by drinking beer." De Bella did not consider someone who only drank beer all day to be drunk. Concerning the morning of the murder, De Bella said that he could not "tell whether he [Kemmler] was sober or not."[49]

Sickmon then tried to cast doubt on the apparent motive offered by the prosecution, and perhaps to suggest that Kemmler was provoked. However, De Bella was a difficult witness to control. He was highly suggestible, yet given to contradictory and often surprising answers. Under cross-examination, De Bella said that he never heard Kemmler say a harsh word to Ziegler, that when they fought, she did all the hollering. De Bella also testified that Kemmler's drinking often angered Ziegler and when it did, she would call him a "rum-done" or hopeless drunk. With some prodding, De Bella told of an all-night "wrangling" between Kemmler and Ziegler in which one or the other, he could not be certain whom, called the other a "son of a gun." When Sickmon asked if those were the exact words used, De Bella, slightly embarrassed, responded, "Yes, son of a bitch."[50] In addition, De Bella said that Kemmler was a habitual drinker, and on the morning of the murder he got up "groggy . . . and that his nerves seemed shaky."[51]

Prosecutor Quinby, unhappy with the way Sickmon had twisted and turned his witness's statements around to support the defense's case, sought to discredit De Bella. Immediately recognizing what Quinby had in mind, Sickmon objected, saying that the district attorney did not have the right to impeach the credibility of his own witness. Judge Childs responded that Sickmon had introduced the issue of Kemmler's drinking in the weeks prior to the slaying and, therefore, the district attorney had the right to reexamine him on this subject. As expected, Quinby tied De Bella up in knots, getting him to back off from many of his previous statements.[52]

On re-cross-examination, Sickmon managed to draw out the fact that De Bella had been kept in jail for the past five weeks as a material witness for the people, but the jurors never heard why the court deemed such a drastic step necessary.[53]

In attempting to bring the prosecution's case to a forceful close, Quinby emphasized that Kemmler had confessed to the crime and that he had done so in a manner that indicated that his crime was premeditated. Next, Quinby called Thomas McMahon, Dennis's younger brother, who also worked for peddler Spang. McMahon's testimony merely added another voice to the several witnesses who had observed Kemmler coming out of his house with blood dripping from his hands.[54] Since no one other than little Ella had actually seen Kemmler kill Ziegler, Quinby, taking no chances, apparently felt it was necessary to pile on the evidence.

Next, Quinby called his final witness, Captain Daniel E. Barry of the Buffalo Police Department, Station Two. Barry explained how he got Kemmler to confess to the crime. He said that when Kemmler arrived at the station house he was sober. He freely admitted to killing Ziegler but, at first, refused to describe the morning's circumstances. He told Barry that he would not make a statement, he did not want a "trial or anything of the kind." After drinking a glass of brandy provided to him by Barry, Kemmler was shown a photograph of Ziegler. When asked if it was a picture of his wife, he responded, "No, that is a picture of Tillie Ziegler of Philadelphia."[55] This, of course, was technically correct since Kemmler and Ziegler were never married.

After another glass of brandy, Kemmler was ready to give a statement. In it, he admitted to intentionally killing Tillie Ziegler with a hatchet. He said that Tillie was always threatening to throw him out and send him back to his wife in Philadelphia. "Tillie was continually talking about my first wife, and telling me that she would make me go back to my old bitch, and that was continually preying on my mind, fearing she would tell me to go."[56]

Quinby was about to read Kemmler's confession to the court when he decided it would be better if he first finished with the witness. Quinby asked Barry to tell the court about a subsequent visit with Kemmler at the Buffalo jail. Barry said that about two days later he visited Kemmler, accompanied by the defendant's brother, Henry, and Ziegler's father, Frederick Tripner.[57] There, Kemmler told them that he killed Tillie because he was jealous of Yella De Bella, that he didn't like the way the two looked at each other, that they were much "too familiar." Kemmler added that he had a bad temper.[58]

Upon cross-examination, Charles S. Hatch, another of Kemmler's attorneys, tried to suggest that Captain Barry deliberately gave his client alcohol in order to induce a confession. Barry seemed aware of where Hatch was going with his questions and he began to parse words, answering no when asked if he had given Kemmler whiskey. Following a delicate exchange, Barry admitted that he gave Kemmler brandy, but he argued that it was not to coax a confession out of him. He said that Kemmler said he felt sick, and this was the reason he offered him a glass of brandy. Continuing, Hatch suggested that since Kemmler could not read or write perhaps he did not understand the statement he signed. Barry would have none of this. He insisted that he did not know that Kemmler was unable to read or write until Kemmler signed his statement by making a mark of X.[59]

These were the days before Miranda warnings. Through his questions Hatch suggested that Kemmler's statement was not entirely voluntary, implying that he was offered something in return. "Did you not tell him it would be better for him to make a statement?" "It is usually your custom . . . to get the offender arrested to make a statement, is it not?" Barry stuck to his story, although he had to admit that he gave Kemmler another glass of brandy after he marked his statement.[60]

After this exchange, Quinby stood up and read aloud Kemmler's statement.

Exhibit No. One

Buffalo, N.Y. March 29, 1889.

STATEMENT OF WILLIAM E. KEMMLER

I am 28 years old; was born in Philadelphia, Pa., and at present reside in the rear part of house No. 526 South Division street, Buffalo, N.Y. About one year and a half ago I left Philadelphia to

come to Buffalo, and a partly divorced woman named Tillie
Ziegler coaxed me to bring her with me; she has a four-year-old
child, and the three of us come to Buffalo, and Tillie and myself
started keeping house as man and wife on Elk street, above Michigan. After living on Elk street a short time we moved to the corner
of Carroll and Heacock streets. About six months ago we moved
to the rear part of house No. 526 South Division street. My business is a peddler of produce, &c., and all winter principally. But-
ter and eggs. My father and mother are both dead. Since I came
to Buffalo I have been drinking considerable, and had more or
less quarrels with this woman Tillie. This morning, March 29th,
'89, between 7 and 8 o'clock, I struck Tillie on the head with a
hatchet—I don't know how many times—with the intention of
killing her. After striking her this morning I went out and told the
woman that lives in the front part of the house—her name is Mrs.
Reid—that I had done it. I then went back to my house and
washed my hands. After doing this, I went out to look for a police
officer to tell him to lock me up for what I had done, and on the
way I stopped into Tom Martin's, corner South Division and
Jefferson streets, to get a glass of beer. I went from Martin's to a
saloon on Swan street, near Jefferson, and had a glass of beer. I
then went back to Martin's and asked him where I could get a
police officer, and told him what I done, and said that I was going
to the police station and tell them what I had done if I could not
find an officer on the way, and while we were talking, an officer
came in and took me from there to my house, and from there to
the station-house. When I got out of bed this morning, I washed
myself and went out and cleaned my horses. While I was cleaning
the horse, Tillie came out to the barn and told the boy she wanted
some eggs, and the boy told me. I then went and got the eggs and
give them to her in a basket out of the wagon. She went in the
house with the eggs. After giving her the eggs, I went to the barn
and fixed the stable door. I then walked into the house with the
hatchet in my hand that I fixed the door with. I first got
acquainted with Tillie on Story street, Philadelphia, when I was
peddling. I used to sell her vegetables, and she got to be a customer. I sold her vegetables quite a few years, and she used to find
fault about her husband, saying that he misused her and she was
always sickly. Some mornings when I called with vegetables she

used to be up stairs in bed, and I used to knock at the door and no one answered, and I opened the door and walked in and hollered, and she said come up. I started up to see what she wanted, then she said to stay down; that she did not want nothing that morning. While calling at her house I had frequent connections with her after moving from this house. This morning, March 29th, '89, we talked over things. After going in the house from the barn, Tillie was cooking the breakfast, and Tillie was continually talking about my first wife, and telling me that she would make me go back to my old bitch, and that was continually preying on my mind, fearing she would tell me to go. This has been going on all winter about my old bitch, and every day that I came here she would say sell out, sell all you got and I was afraid she was going to make me go back to this woman, and I know that I would not live happy with this woman after going away from her, and it made me feel very bad. She used to always get at me about this woman, and I told her to quit that, and I told Tillie to go back again; I will fix you up your same little home you had before you made me take you with me. She would not do that, and Tillie said if I go back, you go back, and sell what you got. I sometimes came home a little full, then she would say to me, I am going home. Then it got so in my mind that I would not care to get her off of my mind of throwing the woman to me. Then I kept coaxing her to go and try and settle the case for me, and try and find out for me if my first wife was married or not, because my wife told me that she had been married to a man named Scofield.

I make the above statement of my own free will.

WILLIAM K. X KEMMLER.

My X marks.

Witness:

Henry J. McKeough,

Daniel F. Barry,

Daniel Dugan.[61]

This closed the case for the prosecution. The court was adjourned for the day.

. . .

On May 8, 1889, at 9:30 a.m., Charles S. Hatch opened the case for the defense. He addressed the jury saying that he was not asking for an acquittal, as "it would be monstrous to turn loose upon society a man with such propensities as are his." Instead, Hatch said he would present the jury with evidence that Kemmler lacked the mental capacity to premeditate a crime. The defendant's mental condition was such that he could not premeditate anything, and therefore could not be guilty of murder in the first degree. He said the evidence would show that the defendant drank to excess, that he once tried to drive his horses over a railroad train, and that he once spent $50 on liquor in a single night.[62]

In an effort to save his life, Hatch told the jury that Kemmler's crime was completely irrational: "There is no explanation in the world [that] will show that his terrible mutilation was a deliberate plan to kill." And, in a peculiar twist of logic designed to show that the defendant lacked control over his actions, he offered the jury this: "If he wanted to murder, one blow would have been enough." Hatch then implied that so long as the defendant's life would be spared, any verdict would be satisfactory, including one that carried a sentence of life imprisonment. He underscored this by pointing to the defendant and saying, "Look at the prisoner. There he sits with hardly any feelings, scarcely any understanding of the terrible position in which he is placed." For his part, Kemmler sat motionless, head bowed, with an air of utter indifference.[63]

It was now official. Kemmler's defense would rely on "alcoholic insanity" to mitigate the gravity of his crime. He could have availed himself of a not-guilty-by-reason-of-insanity plea, but there was no credible evidence to support it. Without overwhelming evidence of mental disease, a jury was not going to return a verdict that relieved Kemmler of criminal responsibility. Alcoholic insanity was his best bet to escape the death penalty. It provided the defense with the best chance of establishing that Kemmler lacked the mental ability to deliberate or premeditate a crime, a necessary element for a first-degree-murder conviction. Of the four types of alcoholic insanity (delirium tremens, alcoholic hallucinosis, alcoholic delusional insanity, and alcoholic dementia), it could be argued that Kemmler suffered from the third type. Alcoholic delusional insanity was defined by paranoid symptoms, "characterized by the slow formation of paranoioid [sic] delusion" caused by chronic alcoholism. To be sure, Kemmler's jealousy of a Tillie and De Bella relationship could be made to fit this category.[64]

The first witness for the defense was Charles A. Spang, "a tall, slim, voluble [man] . . . who [had] a hankering after something to drink."[65] He worked for Kemmler, driving one of his fruit and vegetable wagons. Spang testified that Kemmler was a heavy drinker and that the time he spent with Kemmler was "one continual drunk." Spang said that while on their fruit-peddling route, he and Kemmler would stop at nearly every bar on the way, spending $4 or $5 on drinks and leaving both of them very inebriated at the end of each day's work.[66]

Because Spang's parents believed that Kemmler's drinking habits were rubbing off on him, they made him quit Kemmler's employ. When told that Kemmler had ceased his drinking, Spang returned to work for him. "But in two or three days he started the same old story again." Spang further testified that the day before the murder everyone was drinking, meaning specifically Kemmler, De Bella, and himself. "We were drinking whiskey and playing checkers all afternoon." According to Spang, Billie, as he called Kemmler, got extremely drunk, much more than the others. He "[c]ouldn't walk or talk straight, neither one."[67]

Spang told the court that "Billie" was always "shaky" in the morning, so they would immediately go to get a drink together. On the morning of the murder, "after the deed was committed," Spang saw Kemmler in John Martin's saloon. He testified that Kemmler "was pale as a ghost, and trembling" and his eyes looked "wild." Spang added that Kemmler had looked that way for a couple of weeks. When Sickmon asked him to describe further Kemmler's appearance at Martin's Saloon, he said that Kemmler sat quietly and was unresponsive. According to Spang, Kemmler was frequently despondent and quiet for several hours at a time.[68]

Sickmon then asked Spang about Kemmler's behavior during the two weeks prior to the murder. He was asked if Kemmler's "conduct was rational or irrational." Spang did not know what these words meant. After they were explained to him, he replied: "A little unreasonable; kind of a funny way." When asked under cross-examination by Quinby how many years he had been drinking, Spang was forced to acknowledge that he had been a "drinking man" since he was fourteen years old, or approximately twenty years.[69]

In an effort to discredit one of the defense's star witnesses, Quinby asked Spang if he had been drinking before taking the stand to testify. Spang admitted to having three or four drinks of whiskey that morning. He claimed it was the only thing that kept him "straight." Spang explained, "I am troubled with heart disease and it is the only thing that keeps me quiet." With a sarcastic tone in his voice, Quinby then asked

Spang if he stopped across the street for a drink for his heart disease before coming to court. At this, Sprang got angry and said coldly, "Took it because I needed it." Quinby then asked where Kemmler lived before he moved to 526 South Division Street. Spang replied that he did not know. Quinby quipped, "Well, try to remember—if you are sober enough." Mr. Sickmon then objected to Quinby's remark and Judge Childs chastised him, saying, "That is not proper." Spang would not let it go, however. In a firm voice filled with bluster, he said, "I am as sober as any man in the court room."[70] There is little doubt he meant it.

Seeking to establish that Kemmler retained his mental faculties despite his heavy drinking, Quinby asked Spang how much money they made in the fruit and vegetable peddling business. Spang replied that they made "ten dollars or twelve dollars, some days fifteen dollars," with profits averaging $30 to $40 per week. Spang added that he was drunk every day during the two months prior to the murder and the day before the murder was no different. Spang added that during the afternoon before Tillie's murder, he, Yella, and Billie had played checkers for drinks.[71]

Quinby had established that Kemmler could manage a business even if he drank heavily, and he probably should have quit there. But he asked Spang how Kemmler appeared the morning of the murder. Spang said Kemmler looked "[a]s if he wanted a drink."[72] When asked to be more specific, Spang's ready reply was that he looked "wild." Instead of leaving it alone, Quinby pursued the issue, asking Spang to elaborate on the meaning of the word *wild*. Spang said, "Strange like, bad looking in the eyes—a vacant stare."[73] This was not the impression of Kemmler that the prosecution wanted to leave with the jurors.

Next on the stand was Henry Kemmler, William's younger brother. He testified that thirteen months after William and Tillie left Philadelphia, he married Tillie's sister, Bertha Tripner. While in Philadelphia, William was in a constant drunken state. He always carried a bottle of whiskey in his inside coat pocket. On several occasions he acted "very queer." One morning he even threw himself into a water trough. Another time he was so drunk that he forgot to put any "quiler stretches"—holdback straps—on the horses. At times, William's speech did not make sense. When Sickmon asked the witness what he meant by this, Henry replied, "We couldn't understand what he was talking about. I do not believe he knew himself." Telling Henry to keep his conclusions to himself, Judge Childs ordered the last part of his answer stricken from the record.[74]

Henry further testified that when he and Bertha had visited William and Tillie in Buffalo the previous Christmas, William was drinking excessively. On several occasions Henry tried to help his brother's defense by labeling him crazy with alcohol, but each time Quinby objected and Judge Childs instructed the jury to disregard the witness's statements. The judge also warned Henry to stick to the facts and not to draw conclusions. Ultimately, Henry failed to help the defense. Expecting a much different answer, Sickmon asked Henry how his brother acted during his two-week visit to Buffalo. Surprisingly, Henry said that he did not notice anything peculiar in his brother's behavior.[75]

Although Quinby did not need to cross-examine Henry Kemmler, he did so anyway. The district attorney stuck mostly to factual matters concerning Kemmler's life in Philadelphia. The court learned that prior to Kemmler's departure to Buffalo, the brothers had lived together in the same room for two and a half years, in a boardinghouse at 2807 North Sixth Street. Like his brother, Henry was a huckster. He and William worked together for almost ten years before Billie left with Tillie for Buffalo. Also, Ida Forter, the woman William left behind in Philadelphia, still lived on the corner of Marshall and Somerset Streets.[76]

Mr. Thomas M. Carpenter then testified. He was "a big man who . . . has stuck close to pretty Mrs. Henry Kemmler and her husband during the trial."[77] He was a peddler in Buffalo and "became acquainted with the prisoner last Decoration Day," having met him at the Elk Street Market. When asked to tell the jury what he knew of Kemmler's drinking habits, Carpenter said that he was "not much of a speech maker." In response to direct questions, however, Carpenter did quite well, telling several stories about Kemmler's habitual and reckless drinking. Once, he said, Kemmler was "stupid" drunk and had to be carried home.[78]

Mr. Thomas Martin, the redheaded and freckle-faced owner of the saloon at South Division and Jefferson Streets where Kemmler went for a drink following his murderous attack on Tillie, was recalled by the defense.[79] He testified that in the months leading up to the homicide, Kemmler often sat drinking whiskey at a table in the poolroom at the rear of his saloon. Kemmler would sit there for three, four, or five hours at a time, his head down, and not responding to any questions asked of him. On the morning of the murder, Martin, whose saloon was about two hundred feet from Kemmler's home, testified that the defendant acted differently. He looked from side to side and his eyes looked "wild" and "curious."[80]

District Attorney Quinby objected to Martin's characterization of Kemmler. In response, Judge Childs told Martin to describe Kemmler's behavior without characterizing it. An unsophisticated man, Martin was unable to grasp the critical distinction the judge was asking him to make. A bit annoyed, Martin insisted that he was describing Kemmler's behavior as best he could, the only way he knew how. Realizing he was getting nowhere with the witness, Judge Childs simply ordered the jurors to disregard Martin's conclusions concerning Kemmler's demeanor on the morning of the murder.[81]

Upon cross-examination, Quinby easily managed to get Martin to concede that he did not notice anything peculiar about Kemmler's appearance until Mr. King told him not to serve him a drink because "this man has brained his wife." Martin also admitted that it was King's statement, not Kemmler's appearance, that made him think Kemmler looked wild. When Quinby asked Martin if he could demonstrate to the court exactly how Kemmler appeared, the suddenly submissive saloon owner became confused and despondent.[82]

After Martin left the witness stand, the court was forced to adjourn for the morning. Several defense witnesses failed to answer when their names were read aloud in the courtroom. A few of the newspaper reporters speculated that they probably could be found across the street drinking at the local watering hole.[83]

When the trial resumed in the afternoon, some of Buffalo's most colorful characters appeared before the court. Fellow hucksters, saloon owners, and produce-market workers all testified for the defense regarding Kemmler's drinking habits. His almost exclusively Irish drinking buddies gave vivid and spirited accounts of a life ruined by alcohol, relating one calamity after another. Perhaps none was so inglorious as the time that Kemmler attempted to jump his horse—with its wagon still attached—over an eight-foot wooden fence. Although the courtroom frequently erupted in laughter, Kemmler, through it all, sat in his usual position, his head resting on the table, his eyes fixed in a vacant stare, looking like a man who could use a drink.[84]

In an attempt to rebut this testimony, District Attorney Quinby recalled Kemmler's landlady, Mrs. Reid. She testified that in the six months that Kemmler rented rooms from her, she only saw him drunk twice. Upon cross-examination, however, Quinby muscled her into admitting that the night before the killing, Kemmler was seriously intoxicated. Since Kemmler left early in the morning and often did not return home until dark, Mrs. Reid eventually conceded that she did not see him regularly.[85]

Attempting to demonstrate that no matter what Kemmler's drinking habits might have been, he remained a fully capable businessman, seven merchants from the Elk Street Market were called to testify. They said that Kemmler was a "shrewd buyer" who always examined his goods carefully; that he made his purchases and payments as he went along, never paying too much for his merchandise. Clearly the prosecution was attempting to demonstrate that Kemmler's mind was "well-regulated," that he possessed a clarity of intellect sufficient to allow him to premeditate and think deliberately about committing a murder.[86] In short, all seven said they never saw Kemmler behave out of the ordinary, save for the few instances when he was intoxicated. Upon cross-examination, Sickmon attempted to coax the men into talking more about Kemmler's drinking habits, but he got virtually nowhere. Tales of Kemmler's drinking, more than anything, seemed to evoke a humorous response. For example, in response to Sickmon's standard drinking question, one of the men said, "I have seen him pretty well over the bay." Everyone in the courtroom laughed; everyone, that is, except Sickmon, Judge Childs, and, of course, Kemmler.

The trial had become a battle between the two opposing views of the defendant: Kemmler the clearheaded, shrewd businessman and Kemmler the hopeless, thoughtless drunk.

It was now late afternoon, and Judge Childs adjourned the proceeding for the day.

ON MAY 9, 1889, at about 9:30 a.m., the jurors entered the courtroom with clean-shaven faces, undoubtedly the handiwork of juror Jerome Brown, who was a barber by trade. It was Kemmler's twenty-ninth birthday.

The first witness called for the defense was John Kehoe, a young huckster. He provided additional affirmation that Kemmler was a hard and heavy drinker. He told the court that he witnessed Kemmler spending about $50 on alcohol in a single evening. On that occasion Kemmler was in the wine room of the Olympic Theatre, "drinking and throwing his money" around, buying drinks for all the show girls. He got so drunk he had to be carried out and lifted onto his wagon. Upon cross-examination, Quinby asked Kehoe how often he himself drank. The huckster boasted that he drank thousands of times, as often as he pleased, at every opportunity. When asked if he acted irrational when drunk, Kehoe, ever the braggart, retorted. "Yes, every man does."[87] This was exactly the answer the prosecution needed to accentuate its claim that Kemmler was fully responsible for his crime.

The defense then called John Burkhardt. His testimony was merely more of the same regarding Kemmler's constant drinking, until, that is, he testified that the Monday prior to the murder, Kemmler "acted kind of queer to me." When asked by Sickmon what he meant by "queer," Burkhardt explained that Kemmler acted like he was "under the influence of liquor." Much to the defense counsel's delight, Burkhardt, without prompting, added the legally significant comment that he "did not know what he was doing." Under cross-examination Quinby demanded, "You mean drunk, don't you?" Burkhardt replied pointedly, "No, sir; I don't mean he is drunk, for a drunken man sometimes knows what he's doing." Annoyed, Quinby retorted, "How often do you get drunk?" To which Burkhardt shot back, "How often have you been drunk?" Quinby then sarcastically asked Burkhardt when he first started drinking. The witness just laughed and said, "I have drank when I was a baby." Once again, laughter filled the courtroom. After things calmed down a bit, the witness upheld that Kemmler was "the queerest man I ever see."[88]

Desiring to lay the foundation for the medical experts to follow, Defense Counsel Sickmon recalled Henry Kemmler, the defendant's twenty-five-year-old brother, to the witness stand. Henry told the court that his mother died of consumption about thirteen years before, and his father of "green gog," or gangrene, in a German hospital in Philadelphia. He also said that for as long as he could remember, his father, who owned and operated a small butcher shop, drank intoxicating beverages. Clearly, Sickmon had solicited from Henry the testimony he needed regarding a family history of drinking, but as happened with other defense witnesses, Quinby managed to turn him into a witness for the prosecution. Upon cross-examination, Henry revealed that when he spoke with William at the Buffalo jail, his brother told him that he committed the murder because he was jealous of Tillie and Yella De Bella.[89]

The district attorney was not through with Henry. He asked him if he had not told a reporter for the *Buffalo Express* that his brother was a "good, hard-working man." Not wanting to hurt his brother's case, Henry tried to disavow this statement, first by claiming that he did not remember, and then by asserting that he meant prior to his brother moving to Buffalo.[90]

Next up were the medical experts. Dr. T. D. Crothers was sworn in as a medical expert on behalf of the defense. He was the head physician at Walnut Lodge, a private inebriate asylum in Hartford, Connecticut. In order to establish his credentials as an expert on alcoholic insanity, Sickmon asked him how long he had been the editor of the *Journal of Inebriety*. The doctor told the court that he had been the editor for fourteen or fifteen years, that he specialized in "the study of inebricty and its effects on the brain and morals" of drunkards, and that he had examined between twenty-five hundred and three thousand cases.[91]

Dr. Crothers testified that excessive alcohol abuse rendered the brain defective. When asked how it affects the person morally, the doctor explained that alcohol "destroys a man's character; destroys his veracity,"— the moral power to distinguish right from wrong or good from evil. The doctor continued, "Alcohol destroys a man's will power, breaks it up." Endeavoring to prove that Kemmler's penchant for alcohol was probably inherited, and, thus, not under his control, Sickmon asked Dr. Crothers that if a man's father drank to excess, whether his son inherits a "weak and excitable nervous system and a weak brain." Quinby immediately objected, arguing that "there is no evidence that Kemmler's father drank prior to the time of his birth."[92]

The district attorney's objection sustained, the defense counsel asked the doctor what effect Mrs. Kemmler's pulmonary tuberculosis had on her son. Dr. Crothers replied that there is a close relationship among consumption, drinking, and insanity. "Consumptive parents have inebriate children and inebriate parents have consumptive children." Satisfied, Sickmon asked the physician if a man who began drinking early in life and continued to drink excessively throughout adulthood could have a "normal brain" or "act in a rational manner." As anticipated, Dr. Crothers responded in the negative.[93]

Encouraged by the physician's adroitly tailored answers, Sickmon began to lead his witness. After listening to Sickmon restate some of the foolish things Kemmler did while intoxicated, Dr. Crothers obligingly testified that "such a man [was of] unsound mind and incapable of discriminating as to the quality of his acts." What is more, the doctor told the court that the symptoms of alcoholic insanity included delusions of persecution and spousal infidelity, both of which the defendant undoubtedly possessed. He added that "imagine[d] infidelity" was the most marked symptom exhibited by dipsomaniacs. According to Dr. Crothers, it existed in nearly all cases.[94]

Next, the doctor said he had examined Kemmler and found him to be a "man of weak intellect." He said he noticed that Kemmler had a dilated or "distended pupil," a sure sign of brain disturbance. As further evidence of the effects of alcohol on the human brain, Dr. Crothers testified that Kemmler told him he could not remember the precise circumstances of his crime. "He was hesitating and doubtful, and showed some suspicions." Overall, Dr. Crothers affirmed, Kemmler was a defective degenerate, a man "incapacitated" by alcohol.[95]

Upon cross-examination, Quinby began to dismantle Dr. Crothers's credibility and deconstruct his theory of alcoholic insanity. Through a series of pointed questions, Quinby forced Dr. Crothers to divulge that the Walnut Lodge was a for-profit asylum he founded as a professional business, and the *Journal of Inebriety* was merely part of that enterprise. Dr. Crothers also confessed that he agreed to testify for the defense based solely on a cursory reading of the evidence presented at the coroner's inquest. Perhaps most damaging of all to his credibility was the admission that he had spent less than half an hour examining the defendant and had done it just before the trial resumed, the morning of his testimony.[96]

Next, Quinby asked Dr. Crothers if there was anything distinctive about the disease of alcoholic insanity. The doctor seemed surprised by this question. The best he could answer was to say that there were three types of insanity: mania, dementia, and melancholia. When Quinby pressed his case, asking what kind of insanity he alleged Kemmler suffered from, the doctor backpedaled, claiming that he did not say Kemmler was insane. Realizing Dr. Crothers knew very little about Kemmler, Quinby demanded that the doctor provide details of his examination of the defendant. This forced Dr. Crothers to admit that he did not know whether Kemmler's father commenced drinking before or after he was conceived, or even how often or how much he allegedly drank. Dr. Crothers was aware that Kemmler's mother died of consumption, but he did not know whether it was hereditary.[97]

Picking up his offensive, Quinby then asked Dr. Crothers what possible relevance Kemmler's mother's consumption was in making his diagnosis, since consumption was a disease of the lungs and not the brain. The doctor answered that consumption was "a general disease of the nervous system, as well as the lungs." Mrs. Kemmler's consumption was relevant because hereditary consumption[98] in a parent "produces weakness in the brain power of the child." Quinby was taken aback by this response and asked Dr. Crothers on what authority he made such a statement.

Much to Quinby's chagrin, Dr. Crothers rattled off the names of ten experts on the subject.[99]

Deciding to let go of the subject, Quinby asked if Kemmler had any other diseases. The doctor said that years ago Kemmler had been treated for "ulcers of the genitals," but he did not know if he had been cured.[100] Wisely, Quinby did not follow up on his question. If Kemmler was suffering from syphilis it was clearly in its latent stage and would not have affected his behavior. Mental disorders only appear when syphilis has reached its late or tertiary stage, generally twenty or more years after a person has contracted the bacterium called *Treponema pallidum*.[101]

Quinby continued his assault on the doctor's professional opinions. By insisting that Dr. Crothers defend his diagnosis of alcoholic insanity, Quinby got the doctor to offer some flimsy evidence that he could easily tear apart. When Dr. Crothers said that he knew that Kemmler's brain was defective because of its small size, Quinby asked him how he could tell. "Can you tell the size of a man's brain by looking at his head? Is a small brain necessarily a weak brain?" Then Quinby asked Dr. Crothers if Kemmler's head was smaller than normal, smaller than that of defense counsel Sickmon. When the doctor failed to respond, Quinby told him that Kemmler wore a size 7 hat, exactly average for a man.[102]

UPON FURTHER CROSS-EXAMINATION, Dr. Crothers maintained that Kemmler's entire physiognomy was indicative of a defective intellect. His small head, his eyes close together and set far back, the angle of the arch above the eyes, enlarged ears all are sure signs of a defective brain. Not quite knowing what to make of the doctor's testimony, but keenly aware he needed to put a stop to it, Quinby asked Dr. Crothers, "Are not large ears indicative of benevolence?" "Depends on whose head they are on," the doctor quipped. After the laughter died down, Quinby deadpanned, "What portion of the brain does liquor affect?" "The whole brain," said the doctor.[103]

The spirited back-and-forth between Dr. Crothers and Quinby seemed to delight Kemmler, and he laughed openly. It was the first time Kemmler had shown any emotion other than complete indifference. Once he stopped laughing, Kemmler straightened his back and stared at the jurors. When the district attorney asked Dr. Crothers to look across the courtroom so he might compare Kemmler's countenance with a gentleman's sitting in the front row, Kemmler thought Quinby was talking to him. With a broad smile on his face, the defendant stood up, slightly elevated his chin, and gazed deliberately around the courtroom. This broke the ice for Kemmler, and for the rest of the day he behaved like a "reasonable man."[104]

Hoping to reveal further the superficiality of his medical examination of Kemmler, Quinby asked Dr. Crothers if he was aware of "some other facts" concerning his client. Did he know, for instance, that Kemmler was a successful small businessman who had deftly conducted his occupation right up to the morning of the murder, or that he was considered by "intelligent citizens to be a sharp man of business?" The embarrassed doctor was compelled to answer that he did not and to admit that running a prosperous business could be regarded as evidence of a sound mind. He also had to confess that he did not know that Kemmler was sober the morning he murdered Tillie.[105]

Not only did this line of questioning expose the doctor's ignorance of the basics of Kemmler's life, but it gave the district attorney the opportunity to reiterate the prosecution's interpretation of events. Most significantly, it allowed him to strike at the very heart of the defense's case. In the form of a question, Quinby recounted Kemmler's behavior immediately following his murderous attack. In particular, he retold the story of Kemmler's confession to his landlady, Mary Reid. When asked by Mrs. Reid, "What have you done?" Kemmler replied, "I have killed her." To which she responded, "You don't mean it?" It is here that Kemmler hanged himself. He stated flatly, "Yes, I have killed her, and I am willing to take the rope for it." The last part of the statement was singularly devastating to Kemmler's defense of diminished mental capacity for it demonstrated that he knew the nature and quality of his act; he knew that what he did was wrong and that he would undoubtedly suffer the legal consequences.[106]

Next, Quinby asked Dr. Crothers the difference between a hallucination and a delusion. The doctor explained that a hallucination was a "disorder of the senses," while a delusion is "a disordered fact—disordered impression of the brain." After Dr. Crothers stated that the defendant had never suffered from hallucinations, Quinby asked him if he knew De Bella lived with Kemmler and his paramour. Once more, the doctor had to admit that he did not know a basic fact of Kemmler's life, nor did he know that De Bella's bedroom adjoined the Kemmlers'. Again, the district attorney pounded at the core of the defense's case. If Tillie and Yella were really having an affaire d'amour, then Kemmler had a basis for his jealousy of De Bella. He was not suffering from a "delusion" of spousal infidelity as alleged by the defense.[107]

Returning to the subject of alcoholic insanity, Quinby asked Dr. Crothers, "Are you one of these men who believe that a man may be sane one minute, insane the next minute and sane the next minute?" Dr. Crothers refused to answer the question, believing it unfair. Judge Childs became a little annoyed with the doctor and curtly interpreted: "You mean, not to know." Before ending his cross-examination, Quinby tried to discredit Dr. Crothers by questioning his motives for testifying. He returned to the accusation that the good doctor was more interested in collecting his payment than in discovering the truth. He bluntly suggested his testimony was for sale. "What compensation do you get for coming here? Does it depend on the result of the trial?"[108]

Upon reexamination, Defense Attorney Sickmon endeavored to rehabilitate his witness. He asked the doctor if Kemmler's "syphilitic disease," along with steady drinking, might additionally weaken his mind. The doctor replied that it would "degenerate it still further." Then, hoping to rebuild the delicate edifice of alcoholic insanity the defense had constructed, Sickmon asked Dr. Crothers if Kemmler had the "brain power" to deliberate and premeditate a crime after having been continuously drunk for several days without food. As an expert witness, Dr. Crothers was not limited to recounting the facts of the case; he was allowed to draw his own conclusions. As expected, the doctor told the court that Kemmler did not have the mental capacity to premeditate the murder of his paramour.[109]

Upon re-cross-examination, Quinby asked whether he meant to say that a man who knew enough to get up in the morning, tell a boy to water a horse, repair a door himself, and get eggs at the request of his wife, "don't know what he is doing when he takes a hatchet" and butchers her to death. Without flinching, Dr. Crothers asserted, "Yes, sir, I do. He might not be able to control himself." Quinby quipped, "Because he is mad?" "No, sir. He is demented with the insane impulse." After getting Dr. Crothers to say that an "insane impulse is an uncontrollable impulse," and an uncontrollable impulse is a "morbid criminal propensity," Quinby had the doctor right where he wanted him, for if the doctor knew New York state law on criminal responsibility, he never would have agreed that an "insane impulse" or an "uncontrollable impulse" was the same as a "morbid criminal propensity." Quinby read him the law. "A morbid propensity to commit prohibited acts, existing in the mind of a person who is not shown to have been incapable of knowing the wrongfulness of such acts, forms no defense to a prosecution." Crothers had no choice but to confess that he was unaware of the language of the statute.[110]

On redirect examination, Sickmon gave Dr. Crothers the chance to qualify the meaning of his previous statements. The doctor affirmed his testimony that regardless of the name given to the impulse or its source, it was certainly "induced by a defective brain." On re-cross, Quinby asked the doctor if the brains of all men who drank even small amounts of whiskey were defective. The doctor replied that this was his professional opinion. The expression of such an extreme position did little to enhance his credibility with jurors. According to Dr. Crothers's definition, most—if not all—of those attending the trial had defective brains. Following minor bickering between the district attorney and defense counsel over whether Kemmler's taking little Ella to Mrs. Reid's apartment instead of leaving her at the scene of the brutal murder meant that he knew what he was doing or not, Quinby told Dr. Crothers that he could step down; he had no further questions to put to the doctor. This ended the presentation of witnesses for the defense.[111]

The prosecution countered with two of its own medical experts. The first was Dr. William C. Phelps, a practicing physician and surgeon at the University of Buffalo Medical College. Prior to joining the faculty at the medical college, Dr. Phelps had been Buffalo police surgeon. During his two and a half years on the job, he had attended to more than a hundred insane persons. And he had made "a study of the influence of liquor on the brain."[112]

Dr. Phelps testified that he and Dr. William H. Slacer had examined Kemmler the day after the murder. Before Dr. Phelps could answer any questions regarding his examination, Defense Attorney Sickmon objected to Dr. Phelps testifying, on the grounds that Kemmler was compelled to talk with the doctor and, therefore, forced to give evidence against himself. He also claimed that Dr. Phelps should not be permitted to testify because there existed a legally protected confidential relationship between physician and patient. Furthermore, to permit such testimony was a violation of the defendant's Fifth Amendment right against self-incrimination.[113]

Judge Childs decided to postpone a ruling on Sickmon's objection, allowing Quinby to ask a preliminary question instead. In response to Quinby's question, Dr. Phelps testified that prior to examining the defendant, he and Dr. Slacer told Kemmler that they were physicians sent by the district attorney to inquire into his physical and mental condition. Quinby contended that since Kemmler was aware of the purpose of their visit, a legally protected doctor/patient relationship did not exist. Judge Childs asked Sickmon if he was satisfied with the doctor's answer. Sickmon told the judge that he was not and cited the case of *People v. Ira Stout* as his authority.[114] In this case the New York Supreme Court ruled in 1858 that physicians could not be permitted to disclose any information they acquired treating a prisoner in custody. After consulting the law library and reading what legal scholars had to say about the case, Judge Childs permitted Phelps to continue testifying on behalf of the prosecution.[115]

Dr. Phelps's testimony stood in direct opposition to Dr. Crothers's. In response to Quinby's questions concerning the defendant's countenance, the doctor testified that there was nothing about the size of Kemmler's head that indicated a mental defect or deficiency, nothing about the shape of Kemmler's ears, the "protuberance of the frontal bone" over his eyes, or anything about his stature or overall build or size that indicated "any mental enfeeblement." Quinby then narrated the facts of the crime placed into evidence. Assuming these facts to be true, he asked Dr. Phelps if Kemmler was a person who could distinguish between right and wrong. And, given his "I'll take the rope for it" comment, did the defendant fully realize the nature and consequences of his act?[116]

This was, of course, a nearly direct quote from the McNaughtan Rules, which was the legal test of criminal insanity in New York and most other states at the time.[117] Dr. Phelps answered yes to both parts of the question. Quinby should have ended his questioning there, but he asked what effect a parent's hereditary consumption has on his or her offspring. Reluctantly, Dr. Phelps said that the brain of progeny would not have as much "vitality" as that of a child of healthy parents.[118]

Upon cross-examination, Sickmon asked Dr. Phelps about the amount of money he received for testifying for the prosecution in prior criminal cases. This forced him to admit that he expected to earn a minimum of $100 for his services. This was clearly designed to suggest that, because it was influenced by the promise of remuneration, the doctor's testimony could not be trusted. Left unsaid, however, was the fact that of all the principal participants, Kemmler was the only one not being paid.[119]

More significantly, Sickmon pressured Dr. Phelps into conceding that a man could be sane on some subjects and insane on others. For instance, he could be a good businessman yet suffer from delusions on a particular subject. To be sure, partial insanity, while it did not relieve a defendant of criminal responsibility, did speak directly to the ability to premeditate a crime. In addition, all things being equal, Dr. Phelps conceded, a child whose parents suffered from consumption was more likely to become insane than one whose parents were perfectly healthy. "If they were exposed to the causes of insanity, they would become insane quicker than a person who has strong vital signs." Fortunately for the prosecution, Sickmon never asked the doctors what precisely were the causes of insanity. Presumably, they included excessive and prolonged alcohol consumption.[120]

Next, the district attorney called Dr. William H. Slacer, the second medical expert for the prosecution. Dr. Slacer was a practicing physician and surgeon in Buffalo with admitting privileges at the Sisters of Charity Hospital and the Fitch Accident Hospital for sixteen years. Previously, he was connected with the Erie County Alms House, directing their department for insane paupers. Dr. Slacer told the court that during his six years with the almshouse, he supervised between three hundred and four hundred patients. He explained that on the afternoon of March 30 he had accompanied Dr. Phelps to police headquarters, where they both examined the defendant. He corroborated Dr. Phelps's testimony that, prior to questioning Kemmler, they informed him that they were sent by the district attorney to inquire into his physical and mental health.[121]

Sickmon reentered the same objection as before to the doctor's testimony. Once more he was overruled. However, District Attorney Quinby, perhaps fearing a reversal upon appeal, confined his questions to observations of the defendant in the courtroom and to hypothetical questions. Quinby ran through the same questions regarding Kemmler's countenance that he had put to Dr. Phelps. As expected, Dr. Slacer provided similar answers. The only new piece of evidence he furnished was that Dr. Hubbell, an oculist or ophthalmologist, had made some measurements on Kemmler's eye and found no indication of mental weakness.[122]

Hearing this, Sickmon objected, citing section 834 of the New York Code of Civil Procedure. The code prohibited a physician from disclosing information acquired while attending a patient, if that information was necessary for him to examine or treat the patient.[123] It was similar to the protection afforded a client in talking to his attorney and to a parishioner confessing to his clergyman. Sickmon argued that any information in regard to Kemmler's eye was derived from a medical examination and therefore inadmissible in court. Quinby accepted this objection and dropped the issue.[124]

Once again, the district attorney repeated for his medical witness the now familiar narrative of the morning of the murder, emphasizing the rope comment and his thoughtful concern for four-year-old Ella immediately after the homicide. As anticipated, Dr. Slacer stated that it was his professional medical opinion that the defendant knew the nature and quality of his act and knew that it was wrong. In response to a question by the district attorney, Dr. Slacer told the court that a man could drink liquor for years without destroying his "moral nature or moral perceptions."[125]

Upon cross-examination, Sickmon coerced Dr. Slacer into admitting that constant drinking without food would have a "deleterious effect on the brain" and that intoxicants like alcohol could be a source of insanity. Similarly, the doctor conceded to Sickmon that one could transact his business yet be insane. However, Dr. Slacer would not agree that a three- or four-day drinking binge would necessarily cause permanent brain injury. Nor would he agree that there were aspects of Kemmler's physical appearance, the contour of his body, or the shape of his head that would indicate insanity. He did concede, however, that it was better for a person to eat while drinking heavily.[126]

Aspiring toward a conviction of second-degree murder, Sickmon impelled Dr. Slacer to agree that the morning after drinking heavily a person might be liable to "delusions" and that any "little business troubles" might be "exaggerated in his mind." This was a near brilliant reversal of defense strategy. If Sickmon could not prove that Kemmler was drunk when he committed his crime, perhaps he could prove that sudden absence of alcohol—the withdrawal symptoms themselves—impaired his ability to premeditate a crime. Curiously, this explanation was perhaps closer to the truth. In part, at least, Kemmler killed Tillie not because he was drunk, but because he needed a drink.[127] This theory was tacitly endorsed by the officers at the police station, who offered Kemmler brandy to help clear his mind so he could make a statement, and, of course, by Kemmler himself, who after killing Tillie hurried to Martin's saloon for a drink.[128]

Even though Sickmon did little with this line of argument, the district attorney could not let Dr. Slacer's remarks go unchallenged, for they struck at the core of his case for first-degree murder. Upon redirect, he led Dr. Slacer to say that although a man who is getting over a drunken episode may not have a clear mind, he still knows what he is doing. Far-fetched as it was, the doctor agreed with Quinby's analogy that a hangover does not impair a person's ability to think to any greater extent than does awakening from a sleep or eating a heavy dinner: they can leave a man "not as clear and bright" as he was before.[129]

Upon re-cross, Sickmon persuaded Dr. Slacer to agree that it was possible for a person to do all the things that Kemmler did the morning of the murder and still be insane. Likewise, a person could be under the influence of liquor to such an extent that his intellect was weakened or impaired, and still be capable of carrying out his morning chores. At this, Quinby jumped up and asked the doctor if there was any evidence to support his assertion. The doctor answered no. Hoping to gain control of the final testimony in the trial, Sickmon objected to the form of the question. But it was Quinby who got the last word. The district attorney interrupted and reminded the jury that the doctor had just said that there was no evidence this was the case.[130]

The evidence closed, the court then adjourned until 2:00 p.m.

JUDGE CHILDS began his charge to the jury by recounting the bare essentials of the case and reminding the jury that the defense did not dispute any of them. He informed jurors that unless a homicide was excusable or justifiable—for example, if it was committed by a citizen in self-defense or by a police officer in the line of duty—the criminal law considered the "deliberate and premeditated" killing of a human being to be murder in the first degree. In the absence of "deliberation or premeditation" it was considered murder in the second degree. Manslaughter was the killing of a human being "without a design to effect death."[131]

Judge Childs told the jurymen that there were two separate questions they must consider. First, was the killing under the circumstances detailed in the evidence a criminal act? And, second, if so, what crime was the defendant guilty of "in depriving this woman of her life"? The law said the defendant must understand the nature and quality of his act, and know that it was wrong; otherwise, no criminal act had been committed and the defendant was entitled to an acquittal. If he was responsible for committing a crime, then members of the jury had to decide the nature of that crime. If Kemmler premeditated and deliberated about the murder, then it was in the first degree, Childs explained. If he did not, then it was murder in the second degree. And if Kemmler attacked his paramour in the heat of passion without any intent to kill, he was guilty of the lesser crime of manslaughter. Although voluntary intoxication was no excuse for crime, the judge told the jury they could take it into account. The law allowed drunkenness to be admitted as evidence to show a lack of deliberation or premeditation.[132]

Judge Childs marshaled the evidence for the jurymen in a way that today would be regarded as extraordinarily prejudicial, but at the time was considered perfectly acceptable. He described the bloody murder scene and the extensive wounds to Tillie's face and head, noting that it took seventeen hours for her to die. He also reminded the jurymen of Kemmler's "I'll take the rope" remark and the testimony of two police officers who said that the defendant was sober when they examined him at the station house. Pushing on, Judge Childs strongly suggested that jealousy was the motive for Kemmler's brutal murder of his paramour, and he underscored the defendant's callous disregard for the welfare of his victim—his unwillingness to fetch a doctor until after he had gotten himself a drink.[133]

Thereupon, Judge Childs summarized the evidence for the defense. He said that William Kemmler was a hard drinker, that he often spent $4 or $5 a day on alcohol, and that he had been intoxicated the evening prior to the murder. He recounted testimony from several of Kemmler's drinking companions and noted that the defendant's brother had told the court that Kemmler had been drinking constantly for four months prior to leaving Philadelphia for Buffalo. After reciting the testimony of those who had business dealings with the defendant and considered him intelligent and rational, even shrewd, Judge Childs reminded jurors that Kemmler's mother died of consumption and his father was a hopeless alcoholic.[134]

Next, the judge quoted Dr. Crothers's testimony that "as a rule a man who uses alcohol has a defective brain; it destroys his veracity and power to distinguish right from wrong." In addition, a consumptive mother would produce a child with a defective brain, which in turn would lead that child to covet intoxicating beverages. According to Dr. Crothers, the defendant was probably suffering from alcoholic insanity, causing him to have "delusions of infidelity of wife." Nonetheless, Judge Childs told the jurors, the physicians for the prosecution, Drs. Phelps and Slacer, were of the opinion that at the time of the killing the defendant understood the nature and quality of his act, and he knew it was wrong. He knew it would "expose him to death by hanging."[135]

Having marshaled the evidence, Judge Childs restated the law and the questions facing the jury, this time in a bit more detail, adding that jurors were the "sole judges" of factual questions. "You are to take the law from the court, and . . . the facts from the evidence." He told jurors that the prosecution had the burden of proof, the burden to prove beyond "a reasonable doubt" that the defendant was guilty of the crime of murder as charged. Judge Childs then proceeded to define "reasonable doubt" in the broadest terms possible. He said that if jurors were not fully satisfied of the defendant's guilt, if they were not certain of the facts, "then there is a reasonable doubt and the defendant is entitled to the benefit of that doubt."[136]

Continuing, Judge Childs emphasized the importance of the case. It was important to the defendant because his life is at stake. And it was important to the community because the criminal law must be executed for society to be protected. Striving to encourage jurors to decide the case on its merits, the judge told the jurymen that they had been selected because they were "free from any bias," from any preconceived notions of how the case should be adjudicated, and that they were under oath. Judge Childs urged the jury to deliberate carefully, to apply fearlessly the evidence as presented in court, regardless of the consequences.[137]

Jockeying for an edge, Defense Counsel Sickmon asked the court to once again instruct the jury that for Kemmler to be convicted of first-degree murder, he must have premeditated Tillie's homicide. Sickmon also wanted the court to emphasize that it was extremely important to consider the evidence of intoxication in fixing the degree of murder. When Judge Childs told the jury to weigh carefully all the evidence, and that it was not the province of the court to signify what evidence was the most important, Sickmon took exception. District Attorney Quinby, unwilling to let the defense counsel have the last word, asked the court to remind the jury of the brutal nature of Kemmler's crime. In response, Judge Childs declared that jurors were to consider "every circumstance in the case." The case was closed and the jurors retired for the day at 5:40 p.m. Deliberations began after supper and at 10:00 p.m. they were adjourned until the next morning.[138]

By 9:00 a.m. the courtroom was filled with "morbid curiosity seekers" anxious to learn the fate of William Kemmler. Nearly everyone anticipated a quick verdict. At 9:30 Judge Childs entered, followed by Kemmler and his defense attorneys, Sickmon and Hatch, and the jury. When asked if they had reached a verdict, the foreman of the jury rose from his seat to inform the court that they had not yet come to a decision. He then requested to hear once more the testimony of the McMahon brothers, who were helping around the barn on the morning of the murder, and, especially the medical experts.[139]

Judge Childs recited the evidence anew, but this time with a clear bias toward conviction. Moreover, he instructed jurors not to privilege the doctors' testimony over that of ordinary citizens', calling attention to the manner in which the doctors answered questions put to them: "It might be" and "It would likely." He urged the jury not to ignore their own observations on how people behave, that is, to rely on their own common sense and not to be overly influenced by expert medical testimony. After paraphrasing the McNaughtan Rules, Judge Childs noted that "it was not necessary that a man have the highest intelligence to be amenable to the criminal law." He might still be responsible for his actions even if he was the "victim of the alcoholic habit." He told the jurors that even if alcohol did damage Kemmler's brain, all that was necessary to convict him of first-degree murder was that he "understood the nature and quality of his act, and that he meant to kill. If this were not the law, half the human family would be exempt from the consequences of crime."[140]

At 11:18 a.m. the jury returned with its verdict: "The sky . . . had become overcast, and the gloom of the court-room was in keeping with the scene about to be enacted," as the *Buffalo Courier* reported. The clerk ordered Kemmler to stand and at that instant "there was a sharp clap of thunder . . . rain dashed against the windows, [and] as Kemmler turned to face the jury there was a flash of lightning typical of his coming fate." A reporter for the *Buffalo Courier* observed, "To those inclined to find meaning in such fortuitous occurrences, the moment was 'decidedly prophetic.'" Still, Kemmler remained outwardly calm, even as the jury foreman informed the court that it found him guilty of murder in the first degree. In the hope that just one juror might change his mind, Hatch requested that the jury be polled. Each juror was asked if he agreed with the verdict. All answered in the affirmative. Judge Childs stated that next Tuesday would be the day of sentencing. Kemmler was handcuffed to an officer and escorted back to his cell where he would await sentencing.[141]

On Tuesday, May 14, 1889, the courtroom was again filled to capacity. Word had gotten out that Kemmler would be the first person sentenced under the new Electrical Execution Act. This was to be "a sentence of death in terms never before heard in any court of justice in the world." Kemmler reportedly looked nervous and all too aware of his fate. The clerk ordered him to stand to receive his sentence and the courtroom became quiet. When asked if he had "any just cause to show why sentence should not be pronounced," Kemmler replied, "No, sir." Judge Childs lectured Kemmler briefly, suggesting that he spend his final days seeking God's forgiveness. The court's business was almost finished. All that remained was for Judge Henry Childs to pronounce the sentence.[142]

"The sentence is that for the crime of murder for which you stand convicted that within the week commencing Monday, June 24, 1889, within the walls of Auburn State prison or within the yard or enclosure [*sic*] thereof, you suffer the death punishment by being executed by electricity, as provided by the code of criminal procedure of the State of New York, and that you be removed to and kept in confinement in Auburn State prison until that time." Judge Childs closed with the incantation: "May God have mercy on your soul."[143]

Defense Counsel Hatch then took exception to the sentence on the ground that it was cruel and unusual punishment. Kemmler stood for a moment and then sank to his seat. He was handcuffed to an officer and taken back to his jail cell where he would await transfer to Auburn.[144]

KEMMLER SPENT the next nine days at the Buffalo jail. During this time some of his keepers thought that he was "off his base."[145] His attorney, Hatch, told the *Buffalo Evening News* that Kemmler preferred the electric chair to the rope. "He dreaded the long drop and thought he'd rather be shocked to death." Hatch also said that Kemmler was "moody with flashes of wit about him." When visitors would come to his cell door and ask if he were Kemmler, he would tell them playfully that the fellow over there "near the window" was Kemmler.[146] However, Sheriff Jenkins told the *Buffalo Commercial Advertiser* that since his confinement Kemmler had displayed "little or no emotion." He maintained the same "stolid indifference" that characterized his behavior since his arrest.[147] The *Buffalo Evening News* published the following ditty:

> I used to live in Buffalo,
>   The people knew me well,
> I used to go a-peddling,
>   A plenty did I sell.
> My old clothes were ragged and torn,
>   My shoes wouldn't cover my toes.
> My old hat went flippity flap
> With a schuper to my nose.

I can't sing sing,
I won't sing sing,
I'll tell you the reason why.
I can't sing sing,
I won't sing sing,
For my whistle is getting dry.

Signature Wm. Kemmler

In keeping with the lighthearted spirit of the jingle, the paper noted that Kemmler's whistle was almost always wet, and, if the brutal murder of Tillie Ziegler was not cause enough, Kemmler should be put to death because he was a bad rhymester.[148]

At about six o'clock on May 23, 1889, Kemmler boarded the central train to Auburn. He talked freely along the way about many topics, most notably about crime and punishment, murder and the death penalty. Although much had taken place since Ziegler's murder, Kemmler remained prepared to suffer the ultimate penalty for his crime. Upon his arrival at Auburn prison later the same evening, Kemmler was placed in solitary confinement. Initially he was extremely nervous, unable to sit or sleep, but he soon settled down. He was locked up in a steel cell with "no opportunity of talking to any person save his watchers."[149] He would remain there until his execution.

CHAPTER SIX

Unusually Cruel Punishment

T HE EXECUTION of William Kemmler would not take place for another fourteen and a half months. He still had to make legal history.

On June 11, 1889, his attorney, Charles S. Hatch, filed a writ of habeas corpus appeal to the Cayuga County Court. The writ claimed that Kemmler's sentence imposed cruel and unusual punishment, and therefore was in violation of both the Eighth Amendment of the U.S. Constitution and the New York State Constitution. In Latin, *habeas corpus* literally means "you have the body." A writ of habeas corpus is a judicial order used to compel a person detaining another to appear in court to defend the legality of a prisoner's detention or imprisonment.[1] In this case, Warden Charles Durston was called to appear to defend the legality of Kemmler's imprisonment.

Two weeks later, on June 25, 1889, before Cayuga County judge S. Edwin Day, the attorneys for both Kemmler and the People were present to argue the matter. In the six weeks since sentencing, William Bourke Cockran, the well-known Tammany Hall orator and once and future U.S. congressman, had joined Kemmler's defense team. Although Cockran was just thirty-six years old, he was regarded as one of the best defense attorneys in the nation. In his short career he had represented a wide variety of influential clients, including banks, railroads, steamship lines, and tobacco companies. He was a leading authority on public utilities and had represented "all the important gas and electric companies operating in New York City and Long Island."[2] Indeed, Winston Churchill once confided to Adlai Stevenson that he had based his oratorical style on Cockran's, and according to Churchill's son

(By Roseti, from James McGurrin, *Bourke Cockran* [Scribner's, 1949])

W. Bourke Cockran

Randolph, Cockran acted as both a father figure and mentor to the young Winston.[3]

It is nearly impossible to overstate Cockran's immense intellectual and physical bearing. Born in County Sligo, Ireland, Cockran came to America when he was seventeen. He first enjoyed national acclaim for his rousing speech at the 1884 national Democratic convention in Chicago.[4] His rich Irish brogue and his free-flowing, impromptu speaking style always left a deep impression on his audience, whether it be in the court-room or political stage. And his physical stature only added to the effect: "Who that ever saw it can forget that tall, impressive figure . . . the deep-set eyes, with those curious, curved, almost Oriental eyelids, the powerful nose . . . the noble face . . . [and] above all, the marvelous voice charged with mockery, with passion, [and] always with music."[5]

But why did a man of Cockran's stature take up William Kemmler's case? The answer is simple enough: the money. Cockran was not working for Kemmler but for Westinghouse. It was widely reported that Cockran was retained by George Westinghouse to prevent the use of his generators in an electrical execution. Although Cockran insisted that he was not paid by Westinghouse, the evidence clearly suggests the opposite. Virtu-ally all major newspapers asserted that Cockran was in Westinghouse's employ, and Cockran's reputation was not that of a civic-minded criminal defense attorney. He was one of the most established lawyers in New York, his firm took in six figures annually, and he was not known to have done pro bono work for indigent clients. He had also declined two public posts, that of Corporation Counsel of New York and a nomination for judge for the Court of Common Pleas, both times in favor of the financial rewards of private practice.[6] Although Cockran's rumored payment of $100,000 was most likely inflated by the press, he undoubtedly received a substantial amount of money for his legal services.

Throughout the entire affair, Cockran repeatedly denied having received payment for his involvement in the Kemmler case. Cockran's 1948 biographer pointed to his lifelong opposition to capital punishment in explaining his work on the Kemmler case. When asked who was paying his fees, however, Cockran usually tried to dodge the question. When pressed, he would deny all but a public-spirited motivation. Once, Cockran even claimed that he was motivated by a love of dogs. An "intimate friend" of Cockran told the *Electrical Review* that "Cockran has a remarkably fine dog and he is very fond of him. His wife, too, thinks a great deal of him [the dog]. When the experiments were going forward at Edison's laboratory and at the Columbia College School of Mines, Cockran and his wife used to read of the suffering of poor dogs which were made the victims of science. One night Mrs. Cockran had finished reading a peculiarly graphic account of a dog's suffering, when she said: 'Just think how terrible it would be if they would treat our dog like that.' Cockran sprang up and exclaimed, 'The law providing for execution by electricity is unconstitutional. I'll beat it if I can.' When Kemmler was sentenced he immediately took charge of the case, although he knew he wouldn't get a cent."[7] The *New York Tribune* got closer to the truth. After quoting Cockran as professing "his love for humanity and his desire to prevent an inhumane execution," which he said the electric chair would cause, the paper went on to report, "It leaked out that he was in the pay of the Westinghouse Company."[8]

IN THE WRIT of habeas corpus, Cockran and Charles Hatch offered to prove that Kemmler's sentence was cruel and unusual punishment within the meaning of the New York State Constitution and, therefore, could not be lawfully inflicted. The writ also sought "to establish the facts upon which the Court can pass, as to the character of the penalty."[9] Charles F.

Tabor, the attorney general, presented the People's objection to such an inquiry. He argued that the court did not have the authority to take information and render a decision on the constitutionality of the law. His objection was summarily overruled.[10] Tracy C. Becker, a thirty-four-year-old professor of criminal law and medical jurisprudence at Buffalo Law School, was appointed referee.[11] It was stipulated that the referee's fee and expenses would be paid by the state when taking the testimony of the People, and by the defendant when taking the testimony of the defendant. Becker was paid $20 per day.[12] As is customary, the New York Supreme Court granted a stay of execution pending the results of an evidentiary hearing.

THE EVIDENTIARY HEARING began on Monday, July 9, 1889, in Cockran's office at the Equitable Building in New York City. Cockran opened by saying that certain facts regarding electricity were well understood and generally accepted. He acknowledged that men had been killed by electricity and said he would not attempt to dispute that fact. Deputy Attorney General William Poste, who appeared for the People, concurred and said that the only question was whether killing by electricity was cruel and unusual punishment. Prior to the hearing, both parties had agreed to call members of the Gerry Commission and make their testimony the basis of the investigation. A telegram was sent to Elbridge Gerry asking him to be present. But Gerry was in Newport, Rhode Island, aboard his yacht, *The Electra*—so-called because it was equipped with every electrical device yet obtainable—and at the last minute he telegraphed Becker that he would not attend.[13]

Commissioner Gerry's failure to appear at the hearing was treated lightly by the New York press. *Electrical World* said that Gerry should be excused for "preferring life on his yacht to such an ordeal as he [was] likely to encounter at the hands of Mr. Cockran . . . the great lawyer."[14] As a consequence, Harold Brown was the first witness called on behalf of William Kemmler, the relator.[15] Brown had been in the electrical business for thirteen years, but he had no formal education. Beginning in Chicago as an agent for Edison's Western Electric Company, Brown was in charge of selling Edison's electric pen as well as the early telephone and telegraph at the factory on Kinzie Street in Chicago. He was with Western Electric for three years (from 1876 to 1879), ultimately devising an electromedical apparatus for the use of physicians in the relief of pain.

For the next five years, Brown headed the northwestern division of the Brush Electric Company, both as an electrical expert and salesman. When he publicly inserted himself into the battle of the currents, he had been earning his living for about three years as an independent electrical engineer based in New York City.[16]

Once Brown took his seat in the witness chair, Cockran wasted no time in attacking his credentials.

COCKRAN: What is your occupation?

BROWN: I am an electrical engineer.

COCKRAN: Are you a member of the Institute of Electrical Engineers?

BROWN: No, Sir.

COCKRAN: Have you studied in any institution of learning that is devoted to the teaching of that science?

BROWN: I have been engaged in the electrical business for the past thirteen years, practicing as an electrical engineer before any courses were started in the colleges.

COCKRAN: There is in existence an institute of electrical engineers?

BROWN: Yes, Sir.

COCKRAN: And it is in all respects similar to the Institute of Civil Engineers?

BROWN: Yes, Sir.

COCKRAN: And being a member of the Institute of Electrical Engineers is a badge or evidence of a man's professional standing, just as it is in the case of a civil engineer?

BROWN: That may be.

COCKRAN: I ask you the question, if being a member of the Institute of Electrical Engineers is not a badge or evidence of a man's professional standing, just as it is in the case of a civil engineer?

BROWN: Some of the prominent electricians of this country I believe are not members of any college.

COCKRAN: I am speaking of electrical engineers.

BROWN: I think it is simply a matter of choice whether or not a man joins the society or not.[17]

After allowing Brown to describe his thirteen years of experience in the electrical business, Cockran sought to establish that Brown did not possess any special knowledge.

COCKRAN: You are not a physician?

BROWN: I am not a physician.

COCKRAN: Or a surgeon?

BROWN: No, Sir.[18]

Having exposed Brown's lack of medical knowledge of the precise effects of electricity on the human body, Cockran focused his attention on the electric chair itself. In response to his questions, Brown explained that under contract with General Austin Lathrop, superintendent of state prisons of the state of New York, he had furnished three New York state prisons—Sing Sing, Auburn, and Clinton—with 650-light Westinghouse alternating-current dynamos for the purpose of producing electricity for an execution. The three prisons were provided a Wheatstone Bridge to measure the resistance of the condemned to electricity, a Cardew volt-meter to read the exact pressure of the volts being supplied by Westing-house dynamos, and electrodes to be placed in contact with the skin of the condemned man.[19]

Cockran inquired about how the condemned man would receive the deadly current of electricity. Brown conceded that he had not yet worked out the particulars. In an effort to exploit the possibility of mechanical or human error, Cockran asked why the Wheatstone Bridge was necessary—in other words, why the resistance of a condemned man needed to be measured. Brown explained that each human being has a distinctive resistance to electricity, and that not every person would die from the same degree of electrical shock. Brown went on to explain that there had been much research directed toward ascertaining the average human resistance. At present, it was believed by what Brown called electro-medical men to be approximately 2,500 ohms. The Wheatstone Bridge was necessary to determine the resistance of each person. If a subject's resistance were above average, the amount of electricity would be increased to ensure an adequate supply.[20]

Cockran's strategy was now clear. He would prove that the electric chair was cruel and unusual punishment because it could not be relied upon to produce a quick and painless death. To achieve this end, he would attack the state's witnesses, either on their scientific credentials or their personal or professional agenda. He repeatedly emphasized the general lack of knowledge concerning the effects of electricity on the human body. Finally, by arguing that the Wheatstone Bridge was necessary to individualize electrical executions, he claimed that death would probably not be quick and painless, but long and torturous, and therefore cruel and unusual punishment.

With this strategy in mind, Cockran stepped up his interrogation of Brown regarding the measurement of the average human resistance. He asked Brown from what book he had received such information, if that book was the only source, and what he meant by "electro-medical men." Under this intense questioning, Brown became irritable and his responses seemed only to help Cockran further discredit him. He was unable to give the name of the book, and could only recall that it was a German text. Perhaps most damaging was Brown's inability to list more than two names of prominent electromedical men.[21]

Cockran sensed that Brown was floundering and he moved in for the kill. He began to twist Brown's answers, and to create false contradictions from seemingly logical deductions. Through it all, however, Brown held his own. He stuck to his previous statements, and, whenever possible, corrected Cockran's deliberate misinterpretations. Although Cockran had all the advantages afforded a cross-examiner—and he was among the best—Brown knew much more about electricity than Cockran, and Brown's superior knowledge allowed him to keep Cockran at bay. When he hit a dead end, the aggressive and ever-resourceful Cockran would switch tactics. He began to demand concrete answers to hypothetical questions, asking Brown what he would do if "unforeseen contingencies" arose. What if he encountered "unexpected resistance . . . so that death did not ensue . . . ?" Brown insisted he had planned for any and all contingencies. He was forced to admit, however, that all his knowledge on the killing of human beings by electricity came from eyewitness reports of accidental electrocutions. No human being had ever had his resistance taken prior to suffering death by electrical shock.[22]

Cockran kept trying to get Brown to admit that he did not really know what he was doing. But Brown was equal to the task, annoying Cockran by peppering his answers with examples of how a Westinghouse dynamo had produced accidental deaths. In one case he cited, a fireman, "while stepping in a pool of water," grabbed a Westinghouse wire and was killed instantly. The pressure was less than 1,000 volts and the resistance through his wet shoes was "greater than any resistance" that would be encountered in an electrical execution. As Cockran tried to punch holes in his testimony, asking him about the physical condition of the electrocuted man, Brown assured the court that the Westinghouse dynamos he supplied the state were powerful enough to take the life of even the strongest man.[23]

Getting nowhere, Cockran returned to the question of the average resistance of the human body. He asked Brown if he were not correct in saying that first an executioner must send "a gentle current through the condemned man in order to ascertain what his resistance is." When Brown answered, "Not only that, but to keep a scientific record of the fact," an annoyed Cockran fired back, "Was it a matter of science or was it to kill him? . . . You have supplied the machine by which this man Kemmler is to be killed—is it part of your scheme?"[24]

Having mentioned Kemmler by name, Cockran sought to dramatize the hearing by asking if Brown could supply him with a diagram of the chair.

COCKRAN: And you could give us a drawing of this machine?

BROWN: Yes, Sir. I can give you a [wood] cut.

COCKRAN: Can you furnish us with a drawing which would show the man as he is being delivered over to his fate, what things would be touching him, what machinery would be working, and what things would be surrounding him?

BROWN: Yes, Sir.

THIS WAS A SHREWD MANEUVER, for Cockran knew exactly how the public would react. About two months earlier, on the very same day that William Kemmler became the first person sentenced to die by electricity, the *New York Evening Post* had published an editorial responding to the newly released pictures of the chair. In no uncertain terms, the editors denounced the execution apparatus as a medieval torture chamber: "The intention of the law was to execute a criminal quickly, without the *terrible* scenes which often accompanied hangings." But it was doubtful that those who formulated the law meant for the "victim . . . [to] be seated in a formidable-looking chair, . . . his feet encased in shoes which contain damp sponges, have another sponge placed on his head, and his head clamped down with metallic bands." This is ironic since the lack of "terror-giving paraphernalia" and prolonged preparation was one of the main reasons that legislators voted to replace hanging with electricity.[25]

The first day of the hearing was only sparsely attended. Its whereabouts had been kept secret to the extent that the wrong place had been announced. The next morning, however, was much different as word of the hearing's location had leaked out. Cockran's office at 120 Broadway was filled with lawyers, electrical and medical experts, and reporters from all the major newspapers, periodicals, and wire services. After Referee Becker noted Elbridge Gerry's absence, Brown was recalled to the witness stand.

Brown had endured almost a full day of cross-examination, but Cockran was nowhere near finished with him. Cockran immediately went after him, reminding the court of Brown's limited formal education in electricity. Through his questions, Cockran compelled Brown to respond that he did not have a certificate from any electrical institution. He then proceeded to ask Brown about people he claimed had been accidentally shocked and killed by electrical currents. Cockran intended to discredit Brown by demonstrating that Brown had no direct knowledge of these cases. Brown was forced to respond in the affirmative to the following question: "I understand you now to say that your information was entirely based upon reports or from information which you obtained from bystanders or through information you have been free to consult?"[26]

In an attempt to drive home his point that Brown's knowledge and information could not be relied upon, Cockran asked Brown if he had ever known of a case in which a shock that had been fatal to one person was not fatal to another. Although Brown had compiled a list of ninety-one people who had been killed by electricity, he knew very little about the exact circumstances of their deaths. Cockran got him to admit that many people, including Brown himself, had come in contact with a deadly electrical current and had survived. Nonetheless, Brown maintained that these people did not receive the full force of the current. Cockran wanted to prove that the electric chair's capacity to kill was entirely unreliable. And if it was unreliable, it was cruel and unusual punishment.[27]

Next, Cockran tried to expose the biased nature of Brown's animal experiments. He suggested that Brown's experiments were not designed to determine the comparative danger of direct and alternating current as he maintained, but to demonstrate the inherent danger of alternating current. In response to Cockran's questions, Brown acknowledged that he had made public statements condemning alternating current as dangerous. Prior to his public statements, Brown proudly asserted, it was "held up as being absolutely safe."[28]

At this point, Referee Tracy Becker took over the questioning. He was particularly concerned that Brown might be a proxy for Thomas Edison.

> BECKER: You began your experiments with regard to the effect of the electricity in June or July 1888?

BROWN: On animals: on killing animals.

BECKER: Are you connected in any way with any of the electrical lighting companies?

BROWN: No, Sir.

BECKER: Or have you any connection with the Edison Company?

BROWN: No, Sir.

BECKER: Or Mr. Edison?

BROWN: No, Sir, except a personal acquaintance. I have received a great many favors at his hands; it is entirely a personal friendship.

BECKER: You have made the statement that the alternating current used by one of the electric lighting companies was dangerous and that statement has been attacked?

BROWN: My statement was that the high-tension alternating current, not referring to any company, was dangerous.

BECKER: Was that a current used by any particular company?

BROWN: It is used by several.

BECKER: Mr. Edison is connected with an electrical lighting company?

BROWN: Yes, Sir.

BECKER: And that company uses another current which is not deadly in your judgment; is that so?

BROWN: Yes, Sir.[29]

Once Becker had finished, Cockran resumed his questioning. He wanted to know why Brown took such a keen interest in the comparative danger of direct and alternating current. In response, Brown claimed that he had been attacked by several prominent electricians who read public letters before the Board of Electrical Control portraying him as a man profoundly ignorant of the science of electricity. They also suggested, Brown asserted, that he was owned by the Edison Electric Lighting Company. As a result, his reputation was damaged and the alternating current regained its standing as a perfectly safe current.[30]

Brown claimed that in order to redeem himself and safeguard his reputation, he decided to conduct killing experiments on animals. The experiments were intended to prove that the alternating current was more than merely dangerous—it was lethal. In an attempt to expose Brown's experiments as calculated efforts to damage the Westinghouse Company, Cockran challenged Brown's self-proclaimed philanthropic motives. He made it clear that Brown was associated professionally with Edison, and he introduced as Exhibit No. A a circular in which Brown denounced alternating current as "the executioner's current."[31]

Exploring the relationship between Brown and Edison, Cockran asked about experiments at Edison's laboratory in West Orange, New Jersey. Brown was forced to acknowledge that Edison's chief electrician, Mr. Arthur E. Kennelly, and several other laboratory assistants aided him. When asked about his Columbia College experiment, Brown reluctantly revealed that Mr. Kennelly was again present along with a number of doctors and electricians representing various electric, telegraph, and insurance companies.[32]

To be sure, Edison's involvement made it appear as though Brown's experiments were not conducted for the benefit of science. The attorney for the People, George Quinby, the man who had prosecuted Kemmler, rose to cross-examine Brown. He hoped to remove the veil of uncertainty Cockran had adroitly put in place. He asked Brown if, during the experiments on the dogs, calves, and horses, he had measured their resistance prior to applying the fatal current. This question attempted to demonstrate that Brown and his assistants took precise measurements to ensure the reliability of the results, and that the experiments' results could not have been biased toward proving that alternating current was dangerous. Brown answered that resistance was always measured, adding that the "process of taking resistance" was in no way painful.[33]

Quinby then questioned Brown about the technical aspects of the experiments, providing Brown with the opportunity to tell the court that in nearly all cases "the result was fatal . . . and death was instantaneous." Brown further testified that in all but a few cases, the animals did not exhibit signs of pain. The issue of the speed and painlessness of death by electrocution was clearly at the forefront of the case. Whether death by electrocution was, in fact, quick and painless would be crucial in determining the extent to which it was cruel and unusual punishment.[34]

Brown continued to testify in excruciating detail about the precise methodology of his animal experiments. He explained that he attempted to exterminate one animal with 1,420 volts of direct current and failed. When he applied alternating current of a lesser voltage the animal was successfully killed. This testimony helped to support his contention that alternating current was more lethal than direct current. The final question put to Brown permitted him to testify that as a result of his experiments he felt certain that executing criminals with electrical shock would "produce instantaneous and therefore painless death."[35]

Following Quinby's examination, Deputy Attorney General Poste posed broad questions, allowing Brown to demonstrate his in-depth knowledge of electricity. Brown described the relationship among amperes, volts, and ohms. Ampere is "the current which will flow through one ohm of resistance when an electromotive force of one volt is applied." The relationship among these three units can be expressed by "a calculation based on Ohm's law that the current in amperes equals the electro-motive force in volts divided by the resistance in ohms, and from that equation the value of the third may be obtained when two are given."[36] Few spectators other than electricians could have followed his testimony, but it served to establish Brown's credentials in the eyes of newspaper reporters and the public.

In anticipation of the many witnesses who would testify on behalf of Kemmler, Poste asked Brown about survivors of accidental electrocutions. Brown explained that in cases where men had received electrical shocks that should have killed them but did not, the men simply did not get the full force of the current. He explained that there are several factors affecting the contact the victim has with the current, such as whether his clothing was wet or dry, the condition of the wires, and the duration of the contact. If the conditions were such that full contact with the current was not possible, a man would suffer the blow of a shock, but it would not kill him.[37] The design of the electric chair, it would later be argued, provided optimal conditions to produce a quick and painless death.

After Poste was finished with his questioning, Cockran resumed his cross-examination of Brown. He asked Brown whether his pamphlet, "The Comparative Danger to Life of the Alternating and Continuous Electrical Currents," was published to prove that alternating current was "extremely dangerous" while continuing current was "comparatively harmless." Brown answered squarely that this was his purpose, "to show the superiority of the continuous current" in terms of safety. Cockran then asked Brown if there was a nasty struggle between Westinghouse and Edison over this very issue. Brown admitted that the two giant electric companies were in a colossal struggle to dominate the power industry, but he refused to characterize it as nasty or acrimonious.[38]

In response to a question about how he came to conduct experiments at Edison's laboratory, Brown claimed that it was Edison who approached him. Referring to the laboratory, he said, "I didn't choose it." This was in direct contradiction to the story he told newspapers. Edison would later come to contradict this testimony when he took the stand. Brown was also forced to acknowledge that the back of his pamphlet contained a challenge to Westinghouse to dispute his claims.[39] All this made Brown appear to be anything but impartial.

Next, Cockran directed the hearing's attention to the Wheatstone Bridge and attacked the notion that it gave an accurate measurement of resistance. Cockran's pointed questions were designed to reveal that Brown's belief that the Wheatstone Bridge produced accurate resistance measurements was just that—a belief, "and not a matter of demonstration." Brown insisted that it did produce perfectly accurate measurements. Based on his knowledge of the laws of electricity, he knew this to be true. Besides, the leading authorities believed it to be accurate. This was not enough, Cockran argued, to say positively that the Wheatstone Bridge produced accurate measurements.[40]

When asked what exact effect would be produced if the measurement were incorrect or if the current was simply not strong enough to kill the criminal, Brown skirted the issue. When asked again "what effect it would have upon the human frame," Brown explained that the subject would be "unconscious" to any injury or burns, that he would not feel them. After continued questioning, Brown explained that he knew this because it had happened to him. He further explained that the moistened electrodes used with the electric chair would prevent any burning if any such error should occur.[41]

Cockran was not satisfied. He asked Brown how long it would take for the water to dry up, thereby leaving the body to be burned. Brown responded that "it might require all day." Cockran then inquired as to how much water would be used to moisten the electrode, and Brown, getting a bit ill-tempered, answered that he did not know. This, of course, was the answer Cockran was trying to elicit. It made Brown look like he did not know much about the apparatus he helped design and promote.[42]

After some insistent questioning about the possibility of burning, Brown said that it would be possible to keep the electrodes moist with a constant supply of water through a rubber hose system. Cockran pushed harder, posing a hypothetical script in which the water supply would be exhausted. When asked what would happen if that were to occur, Brown was forced to concede that the skin would smolder and burn if the current were kept on long enough. Cockran then commented sarcastically that according to the electrocution law the current would have to be kept on until the criminal was dead.[43]

Next, Cockran attempted to demonstrate that Brown did not have the slightest idea how electricity kills. He asked him to describe in detail the manner in which electricity destroys human life. Brown could only muster a feeble, if not totally unsatisfactory, answer. He said the heart would fail. When Cockran pressed him to define heart failure, however, Brown could only offer the definition that heart failure was a paralysis of the heart's function, the failure of the heart to respond to "nervous pulsations." When asked if his description was accurate, Brown responded that he assumed it to be so. Cockran then got right to the point. "In the case of a person tied up in a chair, such as a criminal would be, is it possible that a current would inflict great torture without making him unconscious?" Brown was forced to respond that if the current were given at a very low pressure, "it might."[44]

In an effort to rehabilitate Brown, Poste, on redirect, asked him if there would be any problem with keeping the electrodes moist enough to prevent burning. Brown replied that there would be no problem. Poste also asked Brown whether he believed it possible to kill a man by placing the electrodes on the head and feet, the proposed method of execution. Brown responded that he believed that to be possible "beyond any doubt."[45]

Quinby then directed the testimony toward the subject of pain. He asked about the speed of nerve sensations in relation to the velocity of the electric current to be used with the electric chair. This allowed Brown to testify that the velocity of electricity was approximately 1 million times faster than that of human nerve sensations, thus precluding the possibility of the condemned man feeling any pain.[46]

With Brown's testimony, the hearing was adjourned for the day. Brown would later be recalled, but for now his ordeal was over. He had "[borne] up bravely . . . under the volley of conundrums fired at him" by Cockran.[47] On the harsh cross-examination by Cockran that Brown was made to endure, the *Electric World* commented that Brown must have experienced the feelings of a "condemned criminal as he is expected to sit, bound and helpless in the fatal electric chair."[48]

The next day, Franklin L. Pope, a former president of the American Institute of Electrical Engineers and currently an engineer for the Westinghouse Electric Company, was called to testify on behalf of Kemmler. His testimony, for the most part, served to demonstrate that the Wheatstone Bridge could not accurately measure the resistance of living organisms. Pope testified that the Wheatstone Bridge was a perfectly reliable device, but was principally used to measure the resistance of objects such as metallic wires. The resistance of human beings, he explained, varies not only from person to person but from measurement to measurement.

He accounted for this variability by explaining that "the chemical action of the current upon the fluids of the human body" made it impossible to get the same reading every time a measurement was taken. The body is in a constant state of change, and, according to Pope, a measurement taken at one point in time could change the next minute.[49]

Much of Pope's testimony focused on Brown's contention that with a sufficient water supply the body would not be burned during an electrocution. Pope expressed doubt about the burn prevention method Brown proposed. He told the court that if the water used to prevent burning should heat up to the point of boiling temperature, the body would surely burn. "A sufficient current," Pope said, "would carbonize a man's body from one end to the other."[50]

Pope then puzzled most observers when he testified that if alternating current were applied properly, death could be produced free of pain. He later told the court that he did not believe that the voltage necessary to kill a human being had yet been determined, but, in any case, the Westinghouse dynamo supplied by Brown was not powerful enough to kill with certainty. Pope further testified that it was impossible to know definitively if any one voltage would kill every man. When asked about the relevance of Brown's experiments on animals in relation to human beings, Pope was again quite honest. He said that Brown's experiment results revealed very little about human death by electrocution.[51]

Upon cross-examination, Pope was asked why the Westinghouse Company objected to the use of their dynamos in the execution of criminals. His answer was characteristically straightforward: ". . . for the reason the public would naturally suppose that a machine that was used for the express purpose of killing . . . would be an unsafe one to put into commercial use." As an employee of the Westinghouse Company, Pope was asked if he knew of the "employment of any counsel to prevent the use" of Westinghouse's dynamos in an execution. He said that he did not.[52]

Referee Tracy Becker then adjourned the hearing for the day.

At Cockran's request, the hearing was moved to Edison's laboratory in West Orange, New Jersey. The visitors were met at the West Orange train station and driven to the laboratory. They were shown around and allowed to see some of Edison's new inventions. Upon entering the phonograph room they were treated to the sound of a solo cornet. Afterward they went down to a laboratory where a Wheatstone Bridge had been prepared for their use. Referee Becker had heard so much about this instrument that he wanted to see one in operation.[53]

One of Edison's assistants obliged the committee and dipped his hands into a solution containing sulfate of zinc. His resistance measured 1,310 ohms. When Kemmler's lawyer Hatch tried the experiment, his resistance was at first measured at 1,360 ohms, then at 1,800, and finally at 1,390. Next, Hatch sat down and took off one shoe and placed his bare foot on a "felt-covered electrode." Harold Brown applied another electrode to the top of his head, and his resistance was measured at 9,870 ohms. With both bare feet placed on the felt-covered electrode, Hatch's resistance was 8,170 ohms. All of this clearly indicated that a man's resistance would vary greatly depending upon the points of contact. Becker was beginning to understand just how delicate a balance needed to be struck for an electrical current to be strong enough to kill quickly and reliably, yet not be so strong as to burn and disfigure the body.[54]

W. Bourke Cockran did not make the trip to Edison's laboratory. He stayed in New York City to serve as a pallbearer at the funeral of a friend. His assistant, Attorney T. D. Kenneson, however, came to the laboratory accompanied by a young man, John H. Noble. One of Edison's technicians recognized Noble as an employee of the Westinghouse Company and told Brown. Noble was not on the list of invited guests and he was considered to have "overstepped the bounds of courtesy" by entering the laboratory of a rival inventor. Brown became agitated and a "hubbub" broke out, with Brown demanding that Noble be removed. Once Kenneson explained that he had invited him to act as his expert consultant, Referee Becker defended his right to be present, arguing that since the State was represented, the Westinghouse people should also be allowed to witness the experiments on the Wheatstone Bridge. Not having much choice, Edison was gracious and did not object.[55]

The following Monday morning, July 15, 1889, the hearing returned to Cockran's office at the Equitable Building. The next three witnesses — Daniel L. Gibbens,[56] a member of the Board of Electrical Control, John H. Noble, the young employee of the Westinghouse Company who had intruded into Edison's laboratory, and Alexander McAdie, an expert on electric lighting — each testified that there was no way to tell what current would in all cases prove fatal. Gibbens interjected a humorous moment into the otherwise deadly serious hearing. In response to a question by Cockran asking if he knew the average human resistance, Gibbens said, "I don't know. But the greatest amount of human resistance that I ever heard of is shown by the electric light companies who do not want their wires to be put underground."[57]

Commissioner Gibbens then compared the effects of electricity and whiskey on different men. He said that one man might drink a pint of whiskey and get fat on it. Another man might drink a pint of whiskey and die. One man would fight a burglar in his house and perhaps arrest him. Another man at the sight of a burglar would be frightened to death. It would be this way with electricity, said Gibbens. Two men might be affected in vastly different ways.[58]

Charles Tupper, "an excitable gentleman" who ran a well-known restaurant at 225 Eighth Avenue in New York City, was called as a witness on behalf of Kemmler.[59] His dog, Dash, a handsome St. Bernard and Scotch Collie mixed breed, had suffered a severe shock from a fallen street-lighting wire. At first the dog appeared to be dead. When a gentleman from the Manhattan Illuminating Company arrived, he suggested that Tupper dig a hole and place the dog in it. This would draw the electricity out of him. Several hours later the dog began to twitch, and after receiving some alcohol he was revived. At this point in Tupper's testimony Dash was brought into the courtroom for all to see. This demonstration, while seemingly trivial, raised a critical question that spoke to the reliability of the electric chair to kill. Had all of Brown's experiments on animals truly led to death, or could some animals have been revived prior to autopsy? What if a man appeared to die in the electric chair, but only suffered a severe shock?[60]

The following day, Dr. Frederick Peterson, who was the head of the Medico-Legal Committee appointed to carry the electrocution law into effect, was called on behalf of the People. He had been the doctor responsible for dissecting several of the dogs killed during Brown's experiments. His testimony helped restore confidence in Brown's animal experiments. He testified that in his opinion animals "can endure a great deal more than a human being."[61] This statement did not necessarily mean that experiments on animals were proof of what would occur with a human being under the same conditions, but it did imply that if animals could be killed by electrocution, so could people.

Peterson also testified about the velocity of nerve sensation as compared to electricity. His testimony that the speed of electricity was "infinitely greater" than that of human nerve sensation supported prior testimony regarding the possibility of producing painless death by means of electrocution. He went on to say that an alternating electric current—which he testified was twice as strong as the direct current—could be produced by a dynamo to kill instantly without pain or burning.[62]

During cross-examination, Hatch asked about the probability of mistaken death. Peterson testified that during some of the dissections he noticed that, after having cut open the body, the right auricle of the heart continued to beat. Nonetheless, he still considered the dogs to be dead because the beating of the right auricle after death "is not an uncommon thing."[63] To be sure, this hurt the case for the People. It raised the possible specter of a man thought to have been killed by the electric chair undergoing what was essentially a live autopsy.

On July 18, 1889, Elbridge Gerry was called as a witness on behalf of both parties. Although he had been subpoenaed to appear, he insisted that he was not trying to avoid testifying. He had been about two hundred miles from New York City at Martha's Vineyard. He said merely that he could not have come earlier because of previous appointments.[64]

Gerry testified that to familiarize himself with executions, he had witnessed a hanging. After examining the hanging, Gerry said that the commission concluded they could not improve upon the current method of carrying out a hanging, and therefore decided to explore other methods. Of the approximately forty methods in use around the world, Gerry explained, the commission focused on four: the garrote, the guillotine, hypodermic injections, and electricity. Gerry testified that he believed that the garrote was preferable to hanging, but it was too closely identified with Spain and her colonies. The guillotine was inseparable from the horrors of the French Revolution, and the spilling of blood was inconsistent with the temperament of the American people. Finally, Gerry said that the commission rejected lethal injection because it required medical men to participate, and they were adamantly opposed. Also, no suitable poison could be readily found. Vapors of prussic acid might be harmful to an untrained technician, and morphine, which Gerry favored, might not work quickly enough on a criminal who had been addicted to drugs. And besides, it might make death "somewhat agreeable" by ridding an execution of its "terrors." Nothing was left except electricity.[65]

Gerry further testified that execution by electricity was first suggested by Dr. Southwick. Although Gerry admitted that more respondents to his survey preferred to retain hanging than switch to electricity, one expert who favored electrocution was Thomas Edison, a man Gerry considered to be "the greatest electrician of modern times." To that statement Cockran sneered: "Did you consider he was the greatest medical man?" "No," replied Gerry, "the greatest electrician." Cockran continued this line of questioning: "Even then did you consider he was a good authority to tell you what would destroy human life?" Gerry replied firmly, "Yes."[66]

Upon subsequent questioning, Gerry revealed that Southwick's initial plan was not to employ an independent dynamo, but to use the electricity supplied to streetlights by the local power plant. This caused an audible murmur among the spectators. Returning to the subject of Edison, Gerry said that he took the master electrician's word as proof that if applied properly, alternating current would go through the body's vital organs, resulting in a painless and instant death.[67]

By now the testimony had become tiresome and hopelessly repetitive. The court stenographer was probably the only one still listening. Yet Cockran pushed forward. After Dr. Landon Gray, a professor of nervous and mental diseases in the New York Polyclinic Medical School, testified that electricity would not kill every person quickly and painlessly, a series of witnesses testified about their experiences with either alternating-current shocks or lightning strikes. Henry M. Stevens, assistant superintendent of the Boston Electric Light Company, explained that he had received a shock of approximately 1,450 volts; because his heart had ceased to beat, he had been presumed dead. This type of testimony raised two disconcerting questions. Is there a definitive way of ascertaining death? And, could there be a problem of mistaken death with the new method? The testimony of witnesses who had suffered different intensities of electric shock and had likewise suffered different injuries lent support to the contention that people's bodies will not always react in the same ways to electric shock.[68]

Next, Dr. Alfonso D. Rockwell was called as a witness for the People. He testified at great length about human resistance and electrode contact. In his autobiography, published thirty-one years later, Rockwell would claim that it was his testimony that carried the day, even while he complained that he was paid just $100 for his appearance—though that was a tidy sum in those days.[69] Dr. Rockwell explained that if an electrode were improperly placed on the skin, the resistance of the subject would increase, making the electrocution difficult and perhaps painful. He further testified that if electrodes were properly placed, the electric chair could execute human beings instantly and without pain. Correct electrode placement would also assure that the electricity would be diffused throughout the body and its vital organs, producing death as a definite result. The doctor also testified that 1,500 volts of alternating current (which he believed to be more dangerous than direct current) would be sufficient to kill on every occasion. It would, he argued, stop the heart, destroy the body's tissues, and paralyze the nerve centers.[70]

On cross-examination, however, Cockran forced the doctor to admit that because no human being had ever been deliberately executed by an alternating-current shock, his testimony was not based on fact, but on mere speculation. Cockran succeeded in making the doctor look unsure of his knowledge. Dr. Rockwell first testified that Brown's method of burn prevention would not be sufficient. Nonetheless, when pressed on the subject, the doctor changed his mind and answered, "I can see no reason why . . . it can't be done."[71]

On July 23, 1889, Thomas A. Edison was called to testify on behalf of the People. This was the moment everyone had been waiting for. A week earlier, Brown had written to Edison's private secretary, Samuel Insull, asking him to encourage Edison and Kennelly to come testify on behalf of the state. "There has been so much absurdity in the testimony of Mr. Westinghouse's witnesses," Brown wrote, "that Mr. Edison could dispose of by a word."[72]

Edison "had a good-natured smile on his broad countenance as he entered the room. His smooth-shaven face, dressed as he was in black broadcloth and wearing a white tie, gave him the appearance of a benignant clergyman."[73] He was now forty-two years old and hard of hearing. The attorney general had to shout questions at him. Edison testified as to the reliability of the Wheatstone Bridge. He affirmed that it supplied an "absolute measurement" and added that extremely high resistance measurements were simply due to improper electrode contact. According to Edison, no burning would occur if the electrode contact were perfect. When the contact is poor, Edison testified, the resistance increases, which results in very little of the current passing through the body. This small amount of current would be insufficient to kill, but it would certainly cause unspeakable pain.[74]

To execute criminals by electric shock, Edison continued, the alternating current, as opposed to direct current, should be employed, as AC power was much more suited to producing intense shock. In an effort to counter Pope's testimony regarding the carbonization of the body, Edison testified that because the chair would produce instantaneous death, the body would not be carbonized, but would rise slightly in temperature as the current passed through it. Edison conceded that there were varying degrees of human resistance, but that proper electrode placement in conjunction with sufficient current would cancel out any related problems.[75] Upon cross-examination, Cockran did not show any special deference to the Wizard of Menlo Park. Under pressure, Edison displayed an astonishing lack of knowledge for an electrician of his stature. He had to ask Kennelly for help with a simple question from Cockran, who asked him to convert nine degrees centigrade to Fahrenheit. He also did not know if his lab possessed a common Thomson voltmeter, again asking Kennelly for help. As Cockran's questioning became more detailed, the solicitor confused the terms *volts* and *amperes*. Edison, now on safe ground, corrected him, eliciting a snide reply from Cockran: "Our minds are not as brilliant on that as yours."[76]

Cockran and Edison then locked horns over the subject of Harold Brown. Edison refused to help Cockran, who was trying to establish the existence of a close working relationship between Brown and Edison. The legendary inventor insisted that he did not remember writing letters of recommendation for Brown, or even meeting him on any more than several chance occasions. Unfazed, Cockran pressed onward and eventually provoked more of Edison's derision. Clearly, the great inventor was not accustomed to being treated with open contempt.

COCKRAN: Mr. Edison, do you know Mr. Brown pretty well?

EDISON: Fairly well. I have seen him about a dozen times.

COCKRAN: Is that all? Did you ever give him a letter of recommendation to anybody?

EDISON: I don't remember one.

COCKRAN: Do you know Mr. Lathrop, the Superintendent of the State Prisons in this state?

EDISON: I do not think so.

COCKRAN: Did he ever apply to you, or did anybody in his behalf apply to you, to furnish a man who would conduct executions with neatness, skill and dispatch?

EDISON: I do not remember any such man. He might have been down to the laboratory.

COCKRAN: Did you ever give Mr. Brown a letter of recommendation recommending him as a person who could meet these requirements, to anybody? To Mr. Gerry, Mr. Lathrop, or anybody else?

EDISON: I don't remember if I did.

COCKRAN: Have you ever given any jobs of any kind to Mr. Brown?

EDISON: No, Sir.

COCKRAN: Are you sure of that?

EDISON: I don't remember any.

COCKRAN: Mr. Edison, do you remember writing a letter to Mr. Johnson early this year concerning Mr. Brown at all?

EDISON: No, Sir. I have written a great many letters to Mr. Johnson, but I do not know whether I wrote one in reference to Mr. Brown.

COCKRAN: Did you ever write a letter to Mr. Johnson in which the following language occurred; I will give it: "It would be a good job upon which to put Brown to work"?

EDISON: I do not remember it.

COCKRAN: Where had you ever seen Brown before this five or six times you speak of? When did you first become aware of his existence?

EDISON: I think the first time I saw him, he came out there about this very business; to Orange.

COCKRAN: And was he a stranger to you then?

EDISON: I think so; I do not think I had ever seen him before.

COCKRAN: Did he come up and ask you to let him have your laboratory for the purpose of killing dogs?

EDISON: He wanted to try some experiments.

COCKRAN: Are you in the habit of giving your laboratory to everybody that asks you?

EDISON: Yes; sometimes I let them experiment there.

COCKRAN: Might I entertain the hope that I may be allowed myself to go there?

EDISON: Yes, Sir; you can come at any time; I will be glad to see you.[77]

In an effort to expose the antagonism between the two electrical giants, Cockran asked Edison how he felt toward Westinghouse. Edison replied, "I do not dislike Mr. Westinghouse."[78] The two were engaged in a fierce legal battle over patent violations, but Edison was too shrewd a businessman, and too conscious of his reputation, to say anything negative about his rival. He did not want his testimony and his endorsement of the electric chair to be seen as motivated by his struggle with Westinghouse to dominate the electrical power industry.

Next up was Arthur Edwin Kennelly, Edison's chief electrician and a future president of the American Institute of Electrical Engineers. Kennelly's testimony mostly served to buttress Edison's. He defended the reliability of the Wheatstone Bridge and reiterated the fact that burning is solely the result of poor electrode contact. Kennelly, like Edison and several before him, testified that an electrical current could be applied to the human body in accordance with scientific principles to produce instant and painless death in every case.[79]

Cockran then endeavored to discredit Brown and Kennelly's animal experiments. He attacked their research methodology and record keeping, pointing out that Kennelly's dissections were presumably performed after death, but the notes of the doctor who assisted him showed that in a number of cases the right auricle of the heart continued to beat during dissection. To Cockran, this indicated that either Kennelly's notes were inaccurate or that some animals presumed dead were in fact alive. Cockran's questioning revealed that Kennelly had not even been present during the dissections, so he could not really have known if the animals were actually dead.[80]

In a particularly fruitful line of questioning, Cockran demonstrated that Brown's and Kennelly's animal experiments—especially those concerning direct current—had not been carefully and scientifically recorded. This, Cockran proposed, was an effort to conceal the ability of direct current to kill. Viewed against the more detailed records from the inventors' experiments with alternating current, this allegation was designed to soil the professional reputation of both Brown and Kennelly.[81]

Cockran then proposed to Kennelly a scenario in which two dogs had equal resistance. One dog would be killed with a certain intensity of shock. He asked Kennelly if he could be certain the second dog would also die from the same intensity of shock. Kennelly had to respond that one could not be certain. Cockran then posed the same scenario substituting people for the dogs. When asked if both men would be certain to die from the same intensity of shock, once again Kennelly had to respond that he did not know. This lent support to Cockran's contention that no one really knew the effect of electricity on the human body and that all the testimony in favor of the electric chair was based entirely on speculation.[82]

After several more witnesses, Brown was recalled. As requested earlier, he presented a diagram of the electric chair, explaining in detail where the electrodes would be placed and how the human body would be strapped into the chair. During this exhibition, Brown credited Dr. Rockwell for having come up with the idea of connecting an electrode to a brass skull-cap about the size of a soup bowl. When asked by Cockran if there would be anything connected to the chair that would prevent the condemned man's head from moving, Brown explained that although the criminal's head would be free to move, nothing would interrupt the flow or line of the current.[83]

Recalled to the stand on behalf of Kemmler, Dr. Landon Carter Gray testified about a French doctor named D'Arsonval who discovered through experiments that he could kill dogs by means of electric shock and then later revive them through artificial respiration. These dogs had apparently shown no signs of life prior to being revived. Gray further testified that in light of D'Arsonval's ability to revive the seemingly dead dogs, he would not have considered the dogs in Edison's laboratory to have been truly dead.[84]

Subsequently, several physicians testified about autopsies they had performed on victims of electric shock. The physicians testified that without knowing that the deceased had received an electrical shock, they would not have been able to determine the cause of death by examining the body. All he would have known, one doctor testified, was that the death had been a result of "suspicious circumstances." The evidentiary hearing was then adjourned for the day.[85]

In order to accommodate witnesses based in Buffalo, the hearing was moved from New York City to Referee Becker's law offices, Fullerton, Becker & Hazel, at 23 West Eagle Street in Buffalo.

On the thirty-first day of July, at 10:00 a.m., the hearing resumed. Dr. John A. Hoffmeyer, the doctor who had performed the autopsy on the body of Samuel W. Smith, the man Dr. Southwick had seen accidentally electrocuted nine years earlier, admitted that had he not known the cause of death he would not have been able to tell that Smith had been electrocuted.[86] It was this type of testimony that lent credence to Cockran's argument that death by electrocution was still in its experimental stages. If several doctors were unable to determine the cause of death, then death by electrocution was still a mystery; no one knew precisely what it did to the human body to cause death. Its effects could not be definitively predicted.

Dr. George Fell testified next for the People. He was both a surgeon and a physician, the man responsible for exterminating stray dogs in the city of Buffalo. He testified that the Society for the Prevention of Cruelty to Animals recommended electric shock as "the best way for taking the life of a dog" because it was considered to be the most humane method of extinguishing life. In addition, Dr. Fell stated that electricity could most definitely produce instant and painless death in humans.[87]

On cross-examination, however, Fell admitted that he could not say for certain what voltage would kill every human being, acknowledging that this information could not be ascertained without doing human experiments. He still felt, however, that enough accidental deaths had occurred to provide a fairly safe estimation. Once again, Cockran managed to get a witness to admit that knowledge of the electric chair was based solely on speculation and unproved assumptions and not on scientific facts. Next, Fell submitted that if the entire body of the criminal were wet, the current would generally tend to stay on the surface of the body, diverting it away from the vital organs. Therefore, in Kemmler's case, he argued, the points of contact with the electrodes should be kept moist to prevent burning, and the rest of him should be kept dry.[88]

Cockran saw an opening. Following Fell's line of reasoning, Cockran inquired about perspiration. If Kemmler were perspiring profusely, would the current not be diverted? Fell responded, "It might have a slight tendency to." When asked if blood was a better conductor than perspiration, Fell responded with a barely audible, "I don't know." If blood were not as good a conductor as perspiration, the current would possibly remain on the periphery of the body if Kemmler were perspiring heavily. In the final analysis, Cockran was able to get Fell to admit that if perspiration were more saline than blood, it would be a better conductor, although Fell still maintained that if the current were powerful enough, the entire body would be equally charged. After Cockran informed the Buffalo doctor of D'Arsonval's experiments on reviving dogs, Fell had to admit that they had an "effect on [his] conclusions."[89]

The next day, August 1, 1889, Dr. Herman Matzinger, a physician at the state insane asylum, testified briefly that his microscopic examination of a man who had been accidentally electrocuted did not indicate the cause of death. With this, the People rested their case.[90]

Charles Durston, the nominal defendant, was the last witness called on behalf of Kemmler. The Auburn warden testified that the installation of the electric chair along with the accompanying paraphernalia necessary to execute Kemmler was "nearly complete." Under cross-examination by Attorney General Poste, he explained that Harold Brown had furnished the prison with an alternating-current dynamo manufactured by the Westinghouse Company. The only thing that had yet to arrive, and it was not in Brown's contract, was a twenty-eight-light lampboard. Once the dynamo was started, the glowing lights would indicate that electricity was passing through the wires. With a series of explicit questions requiring short answers, Cockran in reexamination forced Durston to concede that he did not plan to pretest the effectiveness of the chair:

> COCKRAN: I don't understand as to what your idea of excellence is; what your idea of efficiency is. How are you going to test the efficiency of the instrument before testing it on Kemmler?
>
> DURSTON: In no way whatever.
>
> COCKRAN: So that, in point of fact, your first test will be the application to Kemmler?
>
> DURSTON: Yes, Sir.[91]

After Cockran made the point that Brown would receive full payment from the state regardless of the outcome of the execution, the attorneys for Kemmler rested their case.

The People decided to reopen their case by recalling Harold Brown. This proved a mistake because it allowed Cockran another opportunity to attack Brown's credibility. Under cross-examination Brown revealed that the experiments at the Edison laboratories testing the reliability of the Wheatstone Bridge to measure the average resistance of humans to electricity were hastily conducted. Because they had an afternoon train to catch, and having wasted most of their scheduled two and a half hours touring the various laboratories or waiting in the phonograph room listening to music, the entire experiment itself was compressed into just "eight or ten minutes."[92] The evidentiary hearing was then closed.

CHAPTER SEVEN

Neither Cruel
nor Unusual Punishment

A FEW DAYS after the close of the evidentiary hearing, Harold Brown arrived in Auburn to help install the electric chair. He was joined by his assistant, Edwin Davis, who would remain an executioner for the next twenty-four years. By the time he retired in 1914, Davis had thrown the switch on 240 condemned criminals. Brown told a *New York Times* reporter that all the machinery was in place, but it had not been "belted up" yet. He had expected to conduct some experiments, but was prevented because several scientists he wanted present were unable to attend. He would return in about two weeks and conduct animal experiments. When asked by a reporter if he was "sanguine" that his machine would "knock the life out of Kemmler," Brown, sounding ill-tempered, said, "All the talk to the contrary is rot."[1]

While the report of the evidentiary hearing was at the printer, Brown's private letters were published in the *New York Sun*, unmasking Edison's involvement in the campaign to label alternating current as dangerous and to smear the reputation of the Westinghouse Company. This destroyed any chance Brown still had of portraying himself as a disinterested, civic-minded electrician concerned only with the safety of the public.[2] Public exposure and condemnation forced Brown to lower his profile. His participation in the debate on the relative safety of AC versus DC now became more of a liability than an asset.

On September 17, 1889, less than a month after Brown's private letters were published, Becker's report was submitted to Judge S. Edwin Day of Cayuga County Court. Then, in a little more than three weeks, on October 9, 1889, Judge Day delivered his opinion. It began by restating the facts of the case: how on March 28, 1889, William Kemmler murdered Matilda Ziegler and was later tried, convicted, and sentenced to death under the new Electrical Execution Act. "The primary inquiry is whether, in a proceeding like the present, the constitutionality of the statute under which the prisoner is detained can be impeached." Judge Day noted that both the Constitution of the United States and that of New York, in almost identical language, prohibited "cruel and inhumane punishment." According to Judge Day, there was no need for the court to concern itself with federal statutes, since both Kemmler's crime and Kemmler's conviction came under state jurisdiction. The judge observed, however, that New York law is so benevolent in its approach to the punishment of criminals that "not even he who cruelly murders can be cruelly punished."[3]

Judge Day then identified the two positions argued in the writ of habeas corpus. The first position, held by William Kemmler, was based on the claim that electrocution was a cruel and unusual punishment, because it "may well result in subjecting its unfortunate victim to the most extreme and protracted vigor and subtlety of cruelty and torture." The second position, put forth by the People, held that death by electricity was "a step forward and in keeping with the scientific progress of the age."[4] Death produced by electrocution would prove to be immediate and painless. The "unsightly and horrifying spectacles" associated with hanging would be forever forestalled. Paradoxical as it may seem, Judge Day continued, the court recognized that both of these positions were "professedly based on the grounds of mercy and humanity."[5]

The question of the constitutionality of the law, Judge Day continued, was of primary importance, because should it be held unconstitutional, not only Kemmler, but all those persons who committed capital offenses since the beginning of the year, may escape punishment. Any law enacted to apply to those cases would be voided as ex post facto. Furthermore, there is no clear legal precedent to follow. As a pioneering case, its final decision is likely to have a lasting effect, not only in New York but across the nation. In this regard, Judge Day noted that electrocution had become law only after much more than the "ordinary consideration and deliberation" by the state legislature. All this must be taken into account before a judge can overrule the collective wisdom of duly elected lawmakers.[6]

The question of the constitutionality of the law was also one of fact. Under section 2031 of the Code of Civil Procedure, the court must examine the facts alleged in the writ of habeas corpus, and discharge the prisoner if no lawful cause can be found for his continual imprisonment. The burden of proof to establish the unconstitutional character of the law, however, must be borne by the prisoner. Since scientific questions were involved, Judge Day noted, electrical and medical testimony from experts was necessary to provide the court with the special knowledge it did not otherwise possess. As expected, however, the expert testimony proved to be "conflicting . . . speculative and hypothetical, for on no person has the experiment yet been tried."[7]

Although the phrase "cruel and unusual punishment" has at least a two-hundred-year history, Judge Day said that he would not attempt to define it. While certain types of executions, like crucifixion, dismemberment, or boiling in oil, would be regarded today as illegal, death itself is not a cruel and unusual punishment. The English Common Law rule that applied in New York State before its constitution was adopted, and thereafter until the Electrical Execution Act of 1888 replaced hanging with electrocution, was fully accepted and never legally challenged. Only if electrocution can be "plainly and beyond reasonable doubt" shown to be slower and more painful than hanging can it be declared unconstitutional.[8]

Judge Day further explained that the constitutionality "of an act is to be presumed." When a statute is presumed to be constitutional, "a clear and substantial conflict" must be found before a court can condemn it. Except in rare cases, the courts should not annul the will of the people as expressed through their legislators. In ambiguous and controversial cases, a statute assailed as unconstitutional should not be declared so by the judiciary. This is especially true when such a grave responsibility falls on the shoulders of a single judge.[9]

After quoting case law and numerous legal citations, Judge Day restated three general principles he had extracted. First, every legislative act carries "a presumption of constitutionality." Second, the judiciary should not annul the law in "doubtful cases." And, third, it is ordinarily best for a single judge to leave constitutional matters to the "deliberation and determination of appellate tribunals." Applying the above criteria, Judge Day ruled that Kemmler had failed to meet his constitutional burden. He did not prove that a "force of electricity sufficient to kill any human being with celerity and certainty when scientifically applied cannot be generated." The most that could be said in favor of Kemmler's petition is that there is a "diversity of opinion."[10]

Judge Day declined to discuss the 1,025 pages of testimony taken at the evidentiary hearing, leaving it for the "intermediate and ultimate courts of appeal." In an unusual move, Judge Day explained that since the Court of Oyer and Terminer passed sentenced on Kemmler, they must have considered the disputed law constitutional. In the interest of "judicial comity" and a "decent respect" for that court, Judge Day ordered Kemmler remanded to Warden Durston's custody in the state prison at Auburn.[11]

As anticipated, Westinghouse's attorneys did not let the matter rest. The following week, on October 16, 1889, Kemmler's attorney, Charles Hatch, appealed Judge Day's decision to the general term of the Supreme Court for the state of New York.

W. Bourke Cockran argued the case for Kemmler before Judges P. J. Baker, J. Dwight, and J. J. Macomber. Cockran alleged that Kemmler had been sentenced to endure a "cruel and unusual punishment," and that "he was deprived of liberty, and threatened with deprivation of life."[12] On December 30, 1889, Judge Dwight delivered the court's opinion affirming Judge Day's decision.

Writing for the court, Judge Dwight began his opinion by recounting the history of the constitutional prohibition against cruel and unusual punishment. According to Dwight, the prohibition originated in the 1689 English Bill of Rights. Reacting to the revolution of 1688, this document avowed "that excessive bail ought not be required, nor excessive fines imposed, nor cruel and unusual punishments inflicted."[13] A century later this provision was incorporated into the U.S. Constitution as the Eighth Amendment. In 1846 New York State inserted this provision into its constitution along with the additional clause: "nor shall witnesses be unreasonably detained." A careful reading of Anglo-American legal history clearly indicates that the provision was intended to restrict the power of trial judges who under the common law had broad discretion that they did not always exercise wisely. It was not intended to limit legislative power.[13]

Dwight then stated that punishments such as "burning at the stake, breaking at the wheel, [or] disemboweling," if reintroduced, would undoubtedly be unconstitutional because they were "repugnant" to the provision of the Constitution prohibiting cruel and unusual punishments. It was common knowledge, Dwight continued, that these punishments were "unusual," and, by common consent, they were considered cruel because they "involve torture and a lingering death." Although electrocution was admittedly unusual, it was by no means commonly considered cruel. On the contrary, the common belief was that death by electricity was probably "instantaneous and without pain."[14]

Noting that the Gerry Commission had recommended changing the method of execution from hanging to electrocution in order to make executions more humane, Dwight disputed the claim that the legislature was "ignorant of the character and effect of the penalty prescribed" by the Electrical Execution Act. In the writ of habeas corpus, Kemmler contended that the New York legislature did not fully understand electricity or how it kills and, therefore, could not have made an informed decision on whether it was cruel and unusual punishment. The court countered this argument by reiterating its belief that, except in extraordinary situations, courts must stand by the decision of the legislative body.[15]

Judge Dwight's court further expressed its faith in the New York legislature by arguing that they did not need to know the exact nature of electricity before enacting a new law. The court likened its ignorance of electrocution to its ignorance of hanging. The court did not have precise knowledge of the force of gravity, but it could still evaluate the relative cruelty of hanging. The question before the court, Dwight argued, was not one of fact, but of law. Even if the court had the authority to inquire into questions of fact, Dwight said, his tribunal believed that under precise conditions, electricity would produce instantaneous and painless death in every instance. About this there could be little doubt.[16]

"If the question here were of the wisdom and advisability of the proposed change in the mode of inflicting the death penalty," then there would be a need for prolonged discussion. The court was confined, however, to the question of the constitutionality of law and "we deem further discussion unnecessary." With that Dwight dismissed the writ of habeas corpus and remanded William Kemmler back to the custody of Warden Durston.[17]

Following Dwight's decision, Cockran petitioned the New York Court of Appeals on February 25, 1889, in the faint hope that it would declare electrocution unconstitutional. The brief that Cockran filed with the Court of Appeals divided his legal arguments into five points.

*Point I: The penalty is one that comes strictly
and literally within the constitutional prohibition.*

Cockran began by stating that both sides had agreed that electrocution was an unusual punishment. No attempt to impose a similar penalty had ever been attempted anywhere in the world. What is more, the phrase "current of electricity" was all but "incomprehensible." No one truly understood it; its "very existence is a matter of dispute." While no scientist could state with certainty what amount or force of electricity would cause immediate death, all agreed "that unless death be instantaneous, the agony inflicted on the condemned would be 'intense beyond description.'"[18]

According to Cockran, Attorney General Charles Tabor claimed that a "scientific application of the penalty would *probably* diminish the dangers of torture and that it would *probably* render death immediate and certain." But this was merely scientific speculation. Harold Brown, the man in charge of the electric chair, had repeatedly altered the execution apparatus. Cockran argued that an electrical execution could not be regarded as anything more than an experiment if the executioner himself was uncertain how to carry it out. Even at this late stage, electrical and medical men were still debating the question of what part of the body—the blood or the nerves—would be the best conductor of electricity. It followed, Cockran argued, that the condemned would be "exposed to a punishment which is certainly unusual, and which may be cruel."[19] In this light, Cockran continued, Kemmler's constitutional rights should protect him from even the most minimal risk of torture.

Besides, the Electrical Execution Act prescribed only that a convict should be killed by "a current sufficient to cause death"—the act did not detail how that current should be applied. This, Cockran argued, was nothing more than "a direction that the convict shall be killed." It was like directing that the convict should be killed by a "force of gravitation." Given such a vague directive, the executioner might lead the prisoner to the edge of a precipice, the top of a high building, or the summit of a rock and "execute him by precipitating him to the ground below." Cockran cautioned that wardens—the men charged with the grave responsibility of carrying out executions—were generally not electrical or scientific men. To leave them with the open-ended instruction to kill with a force about which they knew nothing was "absurd, extravagant and unconstitutional."[20]

Point II: The constitutional inhibition renders any law passed by the legislature in violation of its provision, unconstitutional and void.

Cockran acknowledged that the legislature had the power to prescribe punishments for criminal violations. He maintained, however, that the legislature did not have the authority to violate the Constitution by dictating a punishment that was cruel and unusual: "It may take human life, but it cannot experiment with it." In support of his assertion, Cockran invoked the inviolable words of Thomas Jefferson, who maintained that "in popular governments it was absolutely essential to place limitations upon the powers of a majority, and that true liberty could only be secured by prescribing certain limitations of the governmental power within which a majority would be omnipotent, beyond which they could never encroach."[21]

Cockran continued his critique of the Supreme Court of New York's decision by arguing that "if the legislature of this State should undertake to prescribe for any offense against its laws . . . it would be the duty of the court to pronounce upon such an attempt the condemnation of the Constitution." He implored the court not to base its judgment on the testimony of "any scientist, of any expert, or of any philosopher." Instead, Cockran asserted that the decision should rest on the court's "own knowledge, and its own conscience." If Kemmler's execution was painful and prolonged—and Cockran asserted there was no certainty that it would be otherwise—then the Constitution would have been violated and the court would have been a party to that misdeed.[22]

> *Point III: Public policy and the security of our*
> *constitutional system alike demand the enforcement*
> *of this provision by the judicial department*
> *of the government.*

Anticipating the court's reluctance to interfere in the legislature's decision, Cockran's next tactic was to describe the "dangerous results of all disregard of constitutional limitations," as evidenced both by France during the Reign of Terror and England under the Tudors and Stuarts. Cockran pleaded that the court was not being asked to "deny the legislative department of any function which is essential to the security of society," but that if the courts did not review the constitutionality of the Electrical Execution Act now, they might "for all time come to rob themselves of the power of investigating the character of a punishment." The rack, the thumbscrew, and the boot could be reinstated as punishments, Cockran said, and judges would be powerless to prevent their infliction.[23]

Cockran then asserted a kind of constitutional camel's head in the tent theory: that if one part of the Constitution could be "invaded with impunity," then the entire document was left vulnerable. He said, "The security and permanence of our system of government are imperiled if any portion of them can be destroyed or impaired."[24]

Point IV: It being conceded that the infliction of the penalty would expose the relator to the risk of torture, the sentence is in violation of the Constitution, and, therefore, void.

Absent any disagreement about whether electrocution was unusual, and having already reached consensus that any malfunction would at least make it potentially cruel, Cockran stated that a law that "exposes the Constitution to the risk of invasion" was as invalid as one that directly violated its provisions. Quoting New York case law, Cockran proclaimed that "the constitutional validity of a law is to be tested not by what has been done under it, but what may by its authority be done." If the actual execution of William Kemmler should prove to involve an indisputable torture, and this court denied his application, then future courts would be bound to pronounce this penalty on convicted murderers. No "subsequent developments can change its nature in this respect." Cockran then cautioned the court to play it safe and rule that electrocution was a cruel and unusual punishment. "A mistaken conclusion here would be forever irremediable."[25]

Point V: The act is unconstitutional, because upon its face it provides for the infliction of an unusual penalty.

Cockran now approached the most fundamental point in his syllogism: that the Constitution prohibited any punishment that was either cruel or unusual. Cockran maintained that the Constitution, derived as it was from English common law, "must be held to exclude any fanciful or fantastic experiments with human life, and the infliction of any cruel penalty unknown to the common law." Cockran supported his argument with legal precedents from England, quoting philosophers David Hume and Edmund Burke, among other distinguished thinkers of the Enlightenment. He pointed out the difference between a punishment that was "unusual" and one that was "illegal." Had the Framers of the Constitution intended to ban "illegal" punishments, as some judges have interpreted the amendment, it would have been "absurd," since anything that was illegal would clearly be unconstitutional. Instead, according to Cockran, the words were meant to limit how the "power over life and limb may be exercised."[26]

According to Cockran, English common law defined an "unusual" punishment as one that was "unknown to the common law." It was a particular provision of English law that was held as a solemn contract between the king and the people, and had been held, according to Cockran, "as firmly as any constitutional provision which has been adopted by any of the United States." Summoning the massive weight of history, Cockran reminded the court that prohibitions against cruel and unusual punishments were not only modern declarations of principle, but that they had been "asserted from the earliest time." He asked that the "particular provision now under discussion . . . be construed to mean precisely what it meant when it was presented by the famous Free Convention of 1689 to the Prince of Orange as one of the constitutional rights of the subject."[27]

Cockran closed with a harsh and curt criticism of the court: that it had "ignored" the unusual nature of electrocution. The phrase "cruel and unusual punishment" must be held "to exclude any fanciful or fantastic experiments with human life, and the infliction of any cruel penalty unknown to the common law." The court, Cockran said, had "held the penalty to be unusual on its face, [but] did not think that its unusual character was enough to bring it into conflict with the Constitution." He formally ended his brief on behalf of Kemmler with the words: "The order should be reversed and the relator discharged from custody."[28]

In response to Cockran's brief, Attorney General Charles Tabor filed a brief for the People. After running through the facts of the case, Tabor began a point-by-point rebuttal of Cockran's arguments.

HIS FIRST argument was that the relator's appeal to the constitutional prohibition against the infliction of cruel and unusual punishment must be ineffectual, for two reasons. One, a review of the historic origins of the term "cruel and unusual punishment" reveals that in the preamble to the English Bill of Rights of 1689, the term "illegal and cruel punishments" was used to describe the "tyrannical acts of King James II." Tabor explained that prior to 1790, crimes and their corresponding punishments were not regulated by statute. The common law allowed the severity of punishments to be determined by the presiding judge. In the actual enactment of the Bill of Rights, *"illegal* and cruel punishments" became "cruel and unusual punishments." Therefore, Tabor argued, "a prohibition against cruel and unusual punishments was simply and intentionally a prohibition against the illegal usurpations of the courts of criminal jurisdiction and was not intended as a restriction upon the legislative power."[29]

The Eighth Amendment of the U.S. Constitution, Tabor continued, was not intended as a check upon the states, but upon the federal government. The new constitution granted Congress only limited powers to legislate punishments for criminal transgressions, mainly counterfeiting, piracy, and treason. In reference to New York's constitution, Tabor noted, the legal rule of construction *noscitur a sociis*—"it is known from its associates"—must be governing. The context in which the phrase "cruel and unusual punishments" is found—excessive bail and fines, unreasonable detention of witnesses, etc.—can only be applied to the courts. This was its only practicable interpretation.[30]

TWO, Tabor presented his fallback position. Even if the prohibition against cruel and unusual punishment were misconstrued as binding on the legislature, he argued, the Electrical Execution Act did not violate the constitution. Tabor reiterated the argument that all statutes carry with them a presumption of validity. In interpreting the constitutionality of a law, it is important to discern from the statute itself the intent of the legislature. In this regard, Tabor noted, the legislature appointed the Gerry Commission to determine "the most humane and practical method known to modern science of carrying into effect the sentence of death in capital cases."[31]

It is "common knowledge that, with advancing civilization," people have become increasingly uneasy with the use of the gallows. Hanging was neither quick nor painless. Most "educated humanitarians" considered the gallows a "relic of barbarism." Conversely, electricity was the product of scientific investigation. It is both "swifter than thought" and faster than pain impulses. The use of electricity as a method of execution, Tabor proceeded, was "demanded by the humanitarian spirit of the age." The statute was "prompted by mercy to prevent cruelty." It was not intended as a punishment in addition to death.[32]

While Tabor acknowledged that the electric chair was unusual, he argued that it was not cruel. The Constitution forbids cruel and unusual punishment. If it is only unusual, but not cruel, it is not in violation of the Constitution. *Kemmler*, he said, may have established some small degree of doubt regarding the ability of the electric chair to kill quickly and without pain, but that was insufficient. Surely, he did not prove it cruel and unusual beyond a reasonable doubt. And that, Attorney General Tabor proclaimed, was the required standard of proof.[33]

Tabor then argued that the "words of a statute" should be interpreted in their plain and common understanding. For Kemmler's counsel to make the "astonishing assertion" that there was no science of electricity was plainly absurd. It had no meaning. After presenting a brief lecture on electricity, Tabor quoted Edison's chief electrician, Arthur Kennelly, that the only reason that human resistance appeared to vary so widely was the quality of the electrical contact. The average human resistance is approximately 1,000 ohms with a range of between 400 and 18,000.[34]

Tabor then discussed at length the arguments presented by Cockran during the evidentiary hearing, as well as the testimony given by electrical experts. Tabor maintained that a lethal current could be produced that would kill every time. He reminded the court that the execution apparatus had not yet been tested. If the "Westinghouse dynamo which Mr. Brown offers to the State cannot produce a sufficient voltage of current," then another appliance would be obtained. "The State will not run any risks." Science had established that electricity was the "most merciful [method] that has yet been devised by the humanitarian ingenuity of man."[35]

In support of this argument, Tabor insisted that electricity was most assuredly a science. By the assistance of a "defined law of electricity" the telephone and telegraph had become reliable means of communication, passengers were carried by railroad, and the streets of every city were lighted. For Cockran to claim, Tabor continued, that there was no science of electricity, was to deny that "it has lengthened life; it has mitigated pain; it has furnished new arms to the warrior." In short, the law of electricity is the "law of progress." A great orator like Cockran knew this, Tabor insisted, yet he preferred to talk about "carbonization, mummifications," and the "tortures of the rack."[36]

Finally, in a display of professional integrity, or perhaps to limit the possible damage of an adverse decision, Attorney General Tabor questioned Judge Day's opinion that if the Electrical Execution Act was ruled unconstitutional, then Kemmler and those sentenced to die in the electric chair might escape punishment altogether. Although Tabor understood that the court would be less inclined to rule electrocution unconstitutional if it believed that Kemmler and others would be released, he nonetheless expressed his opinion that Kemmler would be sent back to the trial court for resentencing. Citing case law, the attorney general explained that if a person is lawfully convicted, an affirmed writ of habeas corpus would not result in his release. He would instead be remanded for "proper sentence."[37]

On March 21, 1889, writing for the court, Judge O'Brien of the New York State Court of Appeals affirmed the decision of the general term of the Supreme Court of New York. Cutting right to the heart of the matter, Judge O'Brien stated that the amendments under attack "prescribed no new punishment" for the crime of murder. "The punishment now, as before, is death." All that was changed was the "mode" of carrying out the penalty. It might be said, O'Brien continued, that the infliction of the death penalty, in this enlightened age, entails "some degree of cruelty." But, if it were not absolutely necessary to protect society, it would be abolished. Attorneys for Kemmler contended that electrocution exposed their client to the "possible risk of torture and unnecessary pain," but this argument applied to any "untried method of execution," and if extended to its logical conclusion, would prohibit the "enforcement of the death penalty."[38]

O'Brien then dealt Kemmler's petition a crushing blow. He declared that the entire evidentiary hearing, all 350,000 words of it, were irrelevant to the case. If Kemmler could not make the law appear unconstitutional with legal arguments, he ruled, then the act must stand. No amount of expert testimony could change that. According to O'Brien, expert testimony was "not admissible to show that in carrying out a law enacted by the Legislature some provision of the Constitution may possibly be violated." Besides, O'Brien argued, the court agreed with the lower court that the application of electricity to the "vital parts of the human body," under scientific conditions, would result in "instantaneous, and consequently, painless death."[39]

The New York State Court of Appeals turned down Kemmler's appeal and remanded him to the custody of Warden Durston at Auburn State Prison.

The court handed down another ruling the same day. On February 25, 1890, Charles Sickmon, one of Kemmler's trial lawyers, had appealed the verdict of first-degree murder on four separate grounds. First, the court erred when it allowed witnesses to testify regarding quarrels between the defendant and the deceased, to conversations between them, and to statements made by the defendant to others concerning the desire of the deceased to return to Philadelphia. On this, Judge John C. Gray ruled that such testimony was proper because it tended to show the "existence of motives" and other influences acting upon the defendant's mind. Also, it aided the jury's investigation of the possible causes of events.[40]

Sickmon's second objection was much more substantial. He charged that the court erred in allowing the testimony of Drs. Phelps and Slacer, who had been sent by the district attorney to testify regarding the mental condition of the prisoner. Either this violated the confidentiality of the physician/patient relationship or the prisoner was compelled to furnish evidence against himself. Judge Gray ruled that these physicians were sent only to make a physical and mental evaluation of the prisoner. They were not asked to testify about any conversations they may have had with the prisoner or anything else that transpired in the jail. They merely testified as to his mental condition based on their observations in the jail and in the courtroom.[41]

Kemmler's case, Judge Gray continued, was different from that of *People v. Ira Stout*. In the case of Ira Stout, two physicians had been requested by the coroner to examine the prisoner to determine the extent of his physical injuries. Because they actually treated and prescribed for the prisoner, a physician-patient relationship existed, the same as that existing between attorney and client. Allowing these physicians to testify to the physical injuries sustained by the defendant violated the legally protected physician-patient relationship. Unless medical examinations and consultations are privileged, prisoners who have not yet been convicted of a crime would be compelled to suffer "the consequences of injuries without relief from the medical art."[42] In Kemmler's case there was no such error.[43]

Sickmon's third objection was that the trial court erred in refusing to charge that in deciding the degree of murder Kemmler was guilty of, "evidence of his intoxication must be carefully weighted." In response, Judge Gray said that the trial court did charge that "all evidence" was to be "carefully weighted." It was not "the province" of the court to tell jurors what was important testimony and what was not. No error was committed. The trial judge could have characterized the importance of the evidence for the jury, but he was under no obligation to do so. It is the province of the jury to determine the "conclusiveness" of the evidence bearing on the guilt of the defendant.[44]

Sickmon's fourth objection was that the trial judge erred when upon the jury's request—and after it had retired—he provided additional instructions on deciding whether Kemmler was guilty of first- or second-degree murder. On this, Judge Gray said that the additional instruction covered no new ground and the defense made no objection at the time. A consideration of their "tenor" did not reveal anything that might have prejudiced the jury against the defendant. Indeed, he said, the trial was conducted "most fairly." The jury was told applicable rules of law, and although voluntary intoxication did not make the act "less criminal," it might be taken into account in determining the defendant's "purpose, motive, or intent."[45]

After having covered all of Sickmon's objections, Judge Gray decided to comment on the claim that electrocution was cruel and unusual punishment and, therefore, unconstitutional. "Punishment by death, in a general sense is cruel," but it is not cruel within the "sense and meaning of the Constitution." Electrocution itself was new, of course, but it was not "unusual in the sense that some certainly prolonged or torturous procedure" would be unusual. The appeals court must assume that the law as enacted by the legislature was based on an investigation of the facts, and cannot be overruled upon the grounds of "possibilities and guess work" that something might go terribly wrong. Judge Gray concluded by affirming the judgment of the trial court.[46]

On March 31, 1890, Kemmler was resentenced to die in the electric chair during the week of April 28. He told the court:

> I am ready to die by electricity. I am guilty and must be punished. I am ready to die. I am glad I am not going to be hung. I think it is much better to die by electricity than it is by hanging. It will not give me any pain.[47]

BY LAW the decision on the actual date and time of execution was left to the warden: thus, no one knew for certain what day or hour Kemmler would be executed. All the secrecy only stirred up public curiosity, creating a festive atmosphere in Auburn and surrounding communities. Most people believed that the execution would take place at dawn on Tuesday, the twenty-ninth of April, because District Attorney George Quinby of Erie County, who prosecuted Kemmler, Alfred Southwick, who many called the father of electrical execution, and Dr. Clayton Daniels, who was expected to perform the autopsy, had all arrived in Auburn by late Sunday evening.[48]

The rumor that the execution was at hand was strengthened when word leaked out from the prison that electrician Edwin F. Davis of the firm Noel & Davis in New York City was busy installing the execution apparatus. According to a *New York Times* reporter, the chair no longer resembled the "cumbersome affair" that Harold Brown designed the year before. Although still large with a high back, it had a "comfortable seat" and a footrest that might not be necessary. If it were not for the "broad straps" attached to the sides of the chair, its purpose would not be discernible.[49] Later, it was reported that electricians Charles R. Barnes of Rochester and Harold Brown had tested the appliance and it performed to their satisfaction. Brown, a "shy and nervous" man, refused to talk to reporters, but they learned that he and Barnes would be in the dynamo room at the moment the fatal shock was administered, making certain nothing went wrong. Davis would be the executioner. His pay would be $3 plus the clothes Kemmler would wear for the execution.[50]

Early Monday morning an "inoffensive-looking young man" had entered the Osborne House where most of the witnesses were expected to register as guests. He gave his name as Henry B. Gayley, attracting little attention until a *Times* correspondent recognized him as the chief clerk in the law office of Carter, Hughes & Cravath, legal advisers to the Westinghouse Electrical Company. Gayley was evasive when asked just what business brought him to Auburn. Although he said nothing, he "kept his eyes and ears very wide open." When District Attorney Quinby was told of Gayley's presence, Quinby confronted the young counselor in the hotel lobby and began to interrogate him. This led to speculation that Westinghouse was about to make an eleventh-hour attempt to prevent Kemmler's execution.[51]

Rumors of a stay were confirmed later that morning when Warden Durston rushed into the Osborne House and collected the expectant participants. After a brief meeting in a private room, the men came out "with news written on every line of their faces," although they refused to say anything except "the day would witness important developments." An hour earlier Kemmler's newly retained attorney, Roger Sherman, an ex-assistant U.S. attorney from New York City, had obtained a writ of habeas corpus from Judge J. William Wallace, Circuit Court of the United States for the Northern District of New York. The writ read:

The President of the United States to Charles F. Durston, Warden and Agent of Auburn Prison, Greeting:

We command you that you have the body of William Kemmler, by you imprisoned and detained as it is, together with the time and cause of such imprisonment and detention, by whatever name the said William Kemmler is called or charged, before the Circuit Court of the United States for the Northern District of New York, to be held at Canandaigus on the third Tuesday of June at 10 o'clock A.M.

J. W. Wallace

(Witnessed) Hon. Melville Fuller, Chief Justice.[52]

While at the prison notifying Durston of the writ, Sherman demanded to see Kemmler. Durston denied his request on the grounds that he needed proof that Sherman was Kemmler's counsel. A few hours later Sherman obtained an order from New York Supreme Court judge Dwight commanding him to allow Sherman to consult with his client. However, Sherman never did visit Kemmler. Instead, he presented the court order to Durston and then left on the 3:05 p.m. train, not to New York City, where he said he wanted to go, but to Syracuse. His abrupt departure and the misleading information he gave regarding his destination puzzled most observers, especially since Judge Wallace later confirmed that he hurried the necessary papers through without the state seal so that Sherman could consult with his client while still in Auburn.[53]

Meanwhile, the medical and legal witnesses assembled at the Osborne House, annoyed that they had wasted their time, began pointing fingers. Most agreed that "the spoke had again been put in the wheel of justice" by George Westinghouse. Despite Sherman's repeated and emphatic denials they remained perfectly convinced that he was lying. A *New York Times* reporter apparently followed Sherman onto the Syracuse train. The correspondent confronted Sherman and learned that he was on his way to Utica to file Kemmler's petition for the writ. What the *Times* really wanted to know, however, was who hired him to represent Kemmler. On this topic Sherman remained silent.

"Is Kemmler your only client in this matter?"

"Kemmler is the only client whose name I am permitted to mention," he replied.

"Are the Westinghouse people or any other electrical company responsible for your action in the matter?"

"No, emphatically no," said Mr. Sherman. "I cannot tell you by whom I am employed, but I can say that the Westinghouse Company has nothing to do with it."[54]

Another *Times* reporter located Paul D. Cravath of Carter, Hughes & Cravath, and asked if his law firm had anything to do with the last-minute granting of Kemmler's writ of habeas corpus. At first Cravath was evasive, choosing his words carefully and sprinkling his answers with phrases like, "The first intimation which I received that a habeas corpus had been granted" and "as far as I know the Westinghouse Electric Company has taken no steps." But eventually Cravath came out and stated emphatically: "Neither Mr. Westinghouse nor any of his associates had anything whatever to do with the matter."[55]

Warden Durston did not inform Kemmler of his stay until late Monday afternoon. "I know how he will take it," the warden told reporters. "He will manifest as little concern as one of us would if we should receive word that the sun was shining." When Durston did enter Kemmler's cell he said: "You've got a reprieve." "All right," Kemmler replied casually. "It saves us a good deal of trouble," sighed the warden. "It makes me feel a little easier," Kemmler added. It was generally known that Durston did not want to conduct the first electrical execution, and he seemed as relieved as Kemmler.[56]

The other person who seemed relieved was Harold Brown. It was now certain that an execution would not occur before May 1, 1890, the expiration date on Brown's contract with the state of New York. "My contract . . . simply called for the furnishing and setting up of three machines," Brown told reporters. "My contract provided that I was to personally superintend the running of the dynamo at electrical executions," but only until May 1. "There will be no such execution within that limit now," he continued, "and you may rest assured that I am glad to be relieved of the unpleasant responsibility."[57]

The latest stay of execution unleashed a barrage of criticism directed at George Westinghouse and his company. The *New York Tribune* ran an editorial entitled, "Who is Kemmler's Friend?" in which it attacked Sherman's contention that he only "acted in the interest of humanity." It argued that everyone knew Westinghouse was behind all the delays because he feared an electrical execution with alternating current would be detrimental to his business. The paper was also skeptical of Cravath's claim that Mr. Gayley was in Auburn on business other than Kemmler's impending execution. Nor did the newspaper place much stock in Cockran's assertion that he did not hire Sherman to save Kemmler from the chair.[58]

To this purpose, it published a letter from Cockran to Hatch dated April 2, 1890, indicating that Cockran was pulling the strings regardless of whom was officially representing Kemmler. In the letter, Cockran offered Hatch advice on how to challenge further the constitutionality of electrocution. He told Hatch that the new law transferred the power vested in an elected sheriff to an appointed warden, a clear violation of the common law. Since, Cockran proceeded, this was a "question which concerns the body of the convict," it could be reviewed in habeas corpus.[59] Hatch endeavored to carry out Cockran's suggestion, but Judge Corlett of Cayuga County denied the writ.[60] Judge Blatchford, the supreme court justice presiding over the New York Circuit, also refused to grant the writ on those grounds, but suggested that it be made before the full court.[61] Taking over for Hatch, and following Judge Blatchford's advice, Cockran then appealed the denial to the New York Court of Appeals.[62]

Even Judge Wallace, who had issued the emergency stay of execution, came under fire. Kemmler was scheduled to die during the week of April 27, and Judge Wallace's writ was not issued until April 28. This caused some observers to speculate that Judge Wallace, Attorney Sherman, and Warden Durston were all conspiring to protect Westinghouse's business by preventing Kemmler's execution.[63] In a memorandum released afterward, Judge Wallace argued that the question of cruel and unusual punishment raised by Sherman "could only be finally decided authoritatively by the Supreme Court of the United States." Unless Kemmler's claim that his execution by electricity violated the U.S. Constitution was "so frivolous as not to be worthy of serious discussion," a judge was powerless to refuse it. Even the most "depraved criminal" cannot be denied the opportunity to challenge the "validity of his sentence."[64]

If the writ of habeas corpus had been rejected, Judge Wallace continued, Kemmler would have been denied the opportunity to appeal because he would have been dead. This would have placed New York State authorities in an "awkward dilemma," for a hearing could not have been scheduled before the October session of the U.S. Supreme Court. During this time another prisoner under sentence of death in New York could have applied for a similar writ and had all proceedings against him stopped while he awaited a final determination of his sentence. By prior agreement with Kemmler's counsel, the writ could be promptly brought before the court, and a final result achieved in a more timely fashion.[65]

Just when it appeared that Westinghouse had been defeated in his campaign to prevent Kemmler's execution, Assemblyman Newton M. Curtis introduced a bill to abolish capital punishment. The assemblyman from St. Lawrence County had been submitting his bill for the past seven years, but this time the anti–death penalty sentiment was peculiarly strong, and without debate the assembly passed the bill by a lopsided vote of 74 to 29. General Curtis, a Medal of Honor recipient for his service to the Union army during the Civil War, gave a two-hour speech in favor of abolishing capital punishment. Virtually no one thought the bill had a realistic chance of passing; thus, most lawmakers spent the time in the smoking room. Needless to say, Curtis's sudden call for a vote caught the opposition by surprise and only those in favor of the bill were prepared to speak.[66]

Galen R. Hitt, an Albany Democrat with ties to Tammany Hall, made a passionate speech in favor of abolishing capital punishment. "Humanity calls for it and is revolted at executions," he intoned. Only "two classes of people" still supported the death penalty, he said, "clergymen and prosecuting attorneys." Assemblyman Tompkins added that all the chicanery associated with the Kemmler case was the best argument against capital punishment. The long delays, the arcane legal wrangling, and endless corporate bickering had all turned him against the death penalty. And finally, Assemblyman Blumenthal, referring to the futility of executions, declared that "the worst use to which you can put a man is to kill him."[67]

Having passed the assembly, the bill advanced to the senate. Once there Senator Charles T. Saxton, who as chairman of the assembly's Judiciary Committee two years earlier had introduced the Electrical Execution Act, tried to force a quick vote on the bill by immediately calling for a third reading. He might have succeeded had it not been for Senator Cantor of New York City, who strenuously objected to this highly irregular and clearly political maneuver.[68] As a result, the abolition bill was sent to the Senate Judiciary Committee where Chairman Robertson immediately expressed his support.[69]

Upon its arrival, however, the Curtis bill ran into trouble. On May 6, 1890, the Senate Judiciary Committee held a hearing to consider the Curtis bill. An ex-judge, a minister, a few members of the Society of Friends, and the current editor of the *Albany Law Journal* all testified about the worldwide movement to abolish capital punishment and the possibility of innocent people being put to death. Just one or two people testified against the bill, but the committee voted down the bill by 7 to 2, with only Saxton and Robertson voting in its favor. This killed the bill. The reintroduction of another anti–capital punishment bill would have to wait until the next legislative session.[70]

HAVING EXHAUSTED his state appeals, on May 5, 1890, Roger Sherman filed a petition for an original writ of habeas corpus to the U.S. Supreme Court, on behalf of William Kemmler. The petition asserted that Kemmler's death sentence was in violation of the Fourteenth Amendment of the U.S. Constitution because it abridged his privileges and immunities as a citizen of the United States and deprived him of his life without due process of law.[71]

During Sherman's reading of the motion, he was interrupted several times by justices, reminding him that the Supreme Court was an appellate court and not one of original jurisdiction. This meant that he could not introduce new facts or argue old facts already decided by the trial court. In response, Sherman said that there was an "exceptional emergency" and that the court had ruled in a previous case that it might, "under unusual circumstances," grant original writs.[72]

Justice Brewer asked, "You say there are special circumstances which make an emergency. Has not this man been under sentence of death more than a year?" Sherman responded that there had been "no default" on his part and that this was "no dilatory motion for the purpose of securing a stay of proceedings."[73]

Sherman then proceeded to read his petition. He detailed the legal history of the case, from Kemmler's murder trial to the various stages of appeal. He thereupon argued that "the infliction of death by electricity was a cruel and unusual punishment, and that the limited privileges, or immunities, was by the Eighth Amendment and was by the Fourteenth Amendment extended and protected from encroachments by the State." He further argued that even if the Fourteenth Amendment did not exist, the writ should still be granted on the ground that Kemmler's execution by electricity deprived him of life without "due process of law by inflicting cruel and unusual punishment."[74]

Justice Field proclaimed, "If it is instantaneous it cannot be cruel." Sherman responded that that was an "assumption of fact." Unless the current was applied in the "best scientific manner" there was a probability of "prolonged torture, the most frightful ever devised by man." Anticipating what must have been in Justice Field's mind, Sherman added that he did not petition for a writ of error because it would delay the execution a year or more. This was not his purpose. He argued that "sudden death was more or less cruel."[75]

Moreover, even if electricity offered instant death it was unusual and therefore unconstitutional; and there was a "great deal of testimony" showing that much depended upon the skill of the executioner. This was much different from the old law of hanging, which "prescribed" how and when it should be carried out. Under the new law, the warden was authorized to execute the condemned anytime within a week. The new law also gave the warden, who was not an expert on electricity, the authority to apply "such a current as he thought sufficient to kill." The vagueness of the law left open the real possibility of inflicting the "most searching torture the ingenuity of man has ever devised."[76]

Sherman also said that holding Kemmler in solitary confinement, prohibiting the press from publishing reports of his execution—which deprived him of a safeguard against cruel punishment—and not telling him the exact date of his execution all deprived him of "due process of law." In support of this argument, Sherman cited the case of Medley, "a colored murderer" who was released on writ of habeas corpus on these very grounds. In response, Justice Bradley pointed out that the Medley law was passed after the murder, that it was a prejudicial change, and, hence, it was "ex post facto and void."[77]

Sherman acknowledged that in Kemmler's case the law was not ex post facto, but he nonetheless contended that the new law was a "material and prejudicial change" and by "a parity of reasoning" must be regarded as a violation of due process of law. It was a departure from all "settled law" prior to the Electrical Execution Act.[78]

As soon as Sherman finished speaking, Chief Justice Fuller announced the denial of the writ. He stated that inasmuch as a writ of habeas corpus had been granted by Judge Wallace of the U.S. Circuit Court, "and the case [was] proceeding there, we must deny the application."[79]

Justice Blatchford then suggested that Sherman make application for a writ of error. A writ of error is a writ issued from an appellate court, directed to the judge or judges of a court of record, requiring them to remit to the appellate court the record of an action before them, in which

a final judgment has been entered, in order that examination may be
made of errors alleged to have been committed, and that the judgment
may be reversed, corrected, or affirmed, as the case may require. It is the
commencement of a new suit, not the continuation of the original suit.[80]
To secure the granting of this application, Justice Blatchford advised Sherman that he must show that a "federal question was involved necessarily
and essentially in the conviction and sentence of Kemmler."[81]

Writing in the *Albany Law Journal*, Matthew Hale, a member of the
Gerry Commission, assailed Judge Wallace's granting of the writ of
habeas corpus. He asserted that it was the first time a federal judge had
interfered with the execution of "the final judgment of a State court in a
capital case." Although Judge Wallace had the authority to issue it, the
writ amounted to a "reprieve" and a postponement of Kemmler's execution for at least two months. Hale feared that if this precedent were
sustained, any U.S. circuit or district judge could intervene on the eve of
a state execution and give the convict a "new lease on life." If exercised,
this power would create chaos in our legal system, and should only be
used in unambiguous cases to prevent a "plain violation" of the U.S.
Constitution.[82]

In an accompanying editorial, the editors of the *Albany Law Journal*
expressed their concern that the Kemmler case would become a model
for future death row litigation. "If this precedent is to stand," we shall
enter a "new phase of delay introduced in every capital case in which the
ingenuity of counsel can invent a question or suggest a doubt." Judge Wallace should not have substituted his judgment setting aside "the edict of
eleven State judges." This desperate "scheme of delay" was not hatched
in the interest of justice, but of "trade and rival inventors and manufacturers." The editors regretted the "outrageous course" pursued by Kemmler's counsel. There was never a "shadow of doubt" regarding Kemmler's
execution nor a valid point to debate, only some "powerful corporations
or citizens" who were willing to "defeat the laws of this State" to protect
their business interests.[83]

The editors then criticized the way the Curtis bill was handled. They
charged that the legal maneuvering had caused the legislature to act in
haste. The editors did not blame General Curtis or Senators Saxton and
Robertson; instead, they blamed Westinghouse and Kemmler's attorneys.
The editors maintained that they would like to see the death penalty abolished, but not without proper debate and consideration. They closed by
reiterating their misgivings concerning Judge Wallace's suspension of

Kemmler's execution and expressing concern that other states not follow the judge's legal footsteps.[84]

It is interesting to note Edison's success in avoiding a barrage of negative criticism for his role in the adoption and invention of the electric chair. Even with the publication of Brown's letters revealing Edison's involvement in securing a Westinghouse dynamo and the use of Edison's West Orange laboratories for animal experiments, Edison remained unscathed by the controversy and his reputation as a great man continued intact. Westinghouse, on the other hand, suffered constant and prolonged criticism for attempting to postpone or prevent Kemmler's execution with one of his alternating-current dynamos. Each time there was a delay critics would unleash a new volley of criticism, portraying Westinghouse as more interested in protecting his pecuniary interests than in serving the will of the people.

Westinghouse, of course, denied any involvement in the effort to save Kemmler and defeat electrocution. Not until two days after Kemmler's execution, when he released a letter dated May 7, 1890, did he admit even limited involvement. In the letter he said that the only connection his company had with the legal effort to save Kemmler was a letter he wrote to Mr. Sherman objecting to the use of his Westinghouse dynamo. His objection was based solely on the suitability of a "commercial apparatus not designed or adapted" for use as an execution machine. The machine was too powerful and contained too many devices that could fail and thereby inflict the "greatest possible torture" on the condemned.[85]

Since the dynamo had to run at an "abnormal and dangerous speed to obtain the pressure required," Westinghouse continued, failure was almost certain. In testing the execution apparatus in Auburn the previous week upon a calf, "the indicating instruments connected to the machine were destroyed." In closing, Westinghouse said that a "storage battery with an induction coil and a circuit changer" would be much preferred. A simple apparatus, designed by a skilled electrician and powered by a storage battery, would do the job nicely.[86]

ON MAY 20, 1890, the U.S. Supreme Court heard Sherman's petition on behalf of Kemmler for a writ of error.[87] After presenting a brief background of the case, Sherman basically repeated the argument he proffered in his petition for an original writ of habeas corpus. He argued that Kemmler's electrocution would violate both the "privileges and immunities" and the "due process" clauses of the Fourteenth Amendment. It violated "privileges and immunities" because it required that the condemned be held in solitary confinement, and it prohibited the press from publishing reports of the execution, the best protection a prisoner has against cruel punishment. It violated "due process of law" because the condemned was not told the exact date of his execution.[88]

Sherman essentially argued that the state of New York had placed the responsibility of executing condemned criminals into the hands of "young experimenters" who were "trying to find out whether there are any adequate means of killing a criminal." The science was uncertain, Sherman maintained, and the only certainty was "torture of the most horrible kind." Sherman made an additional, moderately mischievous point. After quoting the statute that said that "the current shall be continued" until the prisoner is dead, he argued that the Westinghouse dynamo chosen for the execution would not produce a continuous current but an intermittent one. Thus, the use of alternating current violated the precise letter of the law. Direct current had to be used. The solemn judges just smiled.[89]

The U.S. Supreme Court refused to decide Kemmler's claim that electrocution was cruel and unusual punishment under the Eighth Amendment. The court held that the Eighth Amendment did not apply to the states and, hence, left "unexamined New York state legislature's conclusion that electrocution produced 'instantaneous, and therefore, painless death.' "[90] The court, however, took the opportunity to present the first full analysis of the "pertinent constitutional provisions based upon a Fourteenth Amendment theory of constitutional law."[91]

Writing for the U.S. Supreme Court, Chief Justice Melville Fuller noted the almost identical wording contained in the Eighth Amendment and the New York state constitution, "but this time in the context of interpreting the Fourteenth Amendment."[92] As New York courts had done before them, the Supreme Court held that the Eighth Amendment did not apply to the states. This would, of course, later be overruled in the case of *Robinson v. California* (1962), which incorporated the Eighth Amendment into the due-process clause of the Fourteenth Amendment.[93] Citing the case of *Wilkerson v. Utah* (1878), and reaching all the way back to the *Commentaries on the Laws of England* of William Blackstone (1765–69),[94] the Supreme Court attempted a definition of the phrase "cruel and unusual punishment"—a definition that would become the standard of cruelty cited in countless court cases for more than a century.[95] In *Kemmler*, Chief Justice Fuller wrote for the Supreme Court:

> Punishments are cruel when they involve torture or a lingering death; but the punishment of death is not cruel, within the meaning of that word as used in the Constitution. It implies there is something inhumane and barbarous, something more than the mere extinguishment of life.[96]

The courts of New York, Fuller wrote, had ruled that even though electrocution was unusual, it was not cruel. Today, *Kemmler* generally represents the proposition that a punishment may be constitutional even if it is unusual, as long as it has a humanitarian purpose and effect.[97] The New York legislature passed the Electrical Execution Act in an effort to "devise a more humane method" of executing criminals. The nation's highest court, Justice Fuller wrote, must presume that legislators knew what they were doing when they enacted this law, and they clearly had the power to decide the method of execution. As a result, the U.S. Supreme Court did not have the authority to reexamine the state legislature's decision. New York did not violate "any title, right, privilege, or immunity" afforded Kemmler under the U.S. Constitution.[98]

Moreover, Justice Fuller continued, the Fourteenth Amendment did not alter the relationship between the state and federal governments, nor that of both governments to their people. Protection of "life, liberty and property, rests primarily with the states, and the amendment furnishes an additional guaranty against" the violation of a citizen's fundamental rights, which the state governments were created to protect. The "privileges and immunities" referred to and protected by the U.S. Constitution are those "arising out of the nature and essential character of the national government." To be sure, the Fourteenth Amendment forbids "any arbitrary deprivation of life, liberty or property." In the administration of criminal justice, it "requires that no different or tougher punishment shall be imposed upon one man than is imposed upon all for like offenses."[99]

The enactment of the electrical execution law was within the "legitimate sphere" of New York's legislative power. The legislature determined that electrical execution would not "inflict cruel and unusual punishment, and its courts have sustained that determination." Following this line of reasoning, the U.S. Supreme Court wrote, "We cannot perceive that the State has thereby abridged the privileges and immunities of the petitioner, or deprived him of due process of law." Concluding, the U.S. Supreme Court said that the "judgment of the highest court" of New York could only be overruled if it committed "an error" so grave that it amounted to a denial of due process of law.[100] The record did not support this determination. With that, Chief Justice Fuller denied Kemmler's application for a writ of error on May 23, 1890. The prisoner was once more remanded to the custody of Warden Durston to await execution.

In an editorial, the *New York Times* expressed its relief that the U.S. Supreme Court had finally spoken on the Kemmler matter. Acknowledging that there was still a side issue to be decided—whether the transfer of execution officials from the sheriffs of counties to the wardens of prisons was constitutional—the *Times* staff declared the case essentially over. It decried the "frivolous points" raised by Kemmler's lawyers and said that the process only brought "popular contempt" upon the courts. It had exposed the courts as powerless to protect themselves against "the efforts of people [Westinghouse] who had no real standing in the court." Large sums of money were spent on "unscrupulous efforts" to prevent Kemmler's execution by electricity. It was "monstrous" to use the courts and the condemned man himself for the purpose of protecting "commercial interest."[101]

If what happened in the courts was bad, the *Times* continued, what happened in the legislature was even worse. The action of the assembly in rushing through a bill abolishing capital punishment without debate while the courts were deliberating on the constitutionality of Kemmler's execution was "probably the most disgraceful exhibition ever made . . . by a legislative body in the civilized world." The only justification legislators had offered was that Assemblyman Curtis was "a good fellow" and popular with his associates. They could not be shown to be more unfit to make laws than if they were shown to have taken a bribe for their vote.[102]

The U.S. Supreme Court, the *Times* continued, was forced to listen to "ridiculous arguments" by lawyers who were paid regardless of the final outcome. Their "real clients," meaning the Westinghouse Company, only wanted to "obstruct or delay" justice and they were willing to "incur any expense to avoid the execution of Kemmler by electricity." Even if Kemmler ended up dying of disease or old age, there would be others condemned under the law. "Every decision in this case is a precedent," the *Times* wrote, and it would be quite useless to fight the whole battle over again, even if the clients were willing to pay and the courts as patient as they had been with Kemmler.[103]

. . .

THE FINAL BARRIER to Kemmler's execution was overcome on June 23, when the New York Court of Appeals ruled on the sheriff's question. Taking over for Hatch, W. Bourke Cockran argued for Kemmler that the legislature did not have the power to deprive the sheriff of Erie County of powers conferred upon him by the Constitution. Cockran submitted that the Electrical Execution Act violated the Constitution because it stripped a duly elected official of the powers vested in him when the Constitution was adopted. By giving the power to confine a prisoner out of the "county of his conviction and in an unauthorized building," Kemmler was being "illegally restrained and deprived of his liberty." Cockran demanded that Warden Durston immediately release Kemmler from custody.[104]

Attorney General Tabor argued for the constitutionality of the new law. He began by reviewing all the previous court rulings. About ten minutes into his presentation, however, Chief Justice Ruger cut him short, declaring that further arguments were unnecessary. The court was ready to render a decision. Chief Justice Ruger said that the New York legislature had the power to make the law in question and the decision of the lower court was affirmed. This ended any chance Kemmler had that his execution might be prevented by the intervention of the courts.[105]

IN THE EARLY MORNING of July 12, 1890, William Kemmler, handcuffed to his keeper and friend, Daniel McNaughton, left Auburn by train for Buffalo. He was to be sentenced to die in the electric chair for the third and final time. Having arrived on the 12:15 p.m. train, the two men were conveyed by closed carriage to the Buffalo jail where they would remain until their 2:15 p.m. appointment at the New York Supreme Court. At about 2:00, McNaughton led Kemmler into Judge Childs's courtroom. Warden Durston, who had taken a later train, arrived a few minutes afterward and took his seat inside the rail. The courtroom was jammed with spectators. Most believed that this would be the last opportunity to see Kemmler alive.[106]

Kemmler was neatly dressed in a gray suit. He carried by his side a "natty black derby hat." His scrawny brown beard and mustache had grown thicker during his long confinement. They were neat and fashionably trimmed, his hair tidily arranged. During his trial, he appeared "dull, stupid, and besotted," scarcely aware of what was happening around him.

Today, "he was alert, mindful of his surroundings, as if his faculties had been awakened into life." According to the *Buffalo Evening News*, he displayed a "native resoluteness."[107]

"Mr. District Attorney," said Judge Childs, "have you any business before the court?" Assistant District Attorney Marcy responded, "If the Court please, we desire to move sentence on William Kemmler." After having been asked to rise, Kemmler stood up and straightened his vest. Placing his hands behind his back, Kemmler glanced at Judge Childs. There was a "humble, deprecatory look" on his face. Judge Childs informed Kemmler that all his appeals had failed, that he was at the end of the line. "Yes, sir," Kemmler said respectfully. "Have you anything to say why a time should not be fixed for carrying out the sentence previously passed upon you?" Judge Childs inquired. After a brief pause, Kemmler said: "No, sir."[108]

Speaking in a low, faltering voice, Judge Childs told Kemmler that he hoped that the "long delay" had given him sufficient time to contemplate the "enormity" of his crime and the "justice of his conviction." He then

(New York World, 7 August 1890)

Kemmler in his cell

informed Kemmler that the sentence of death would be carried out during the week of August 4, 1890. Judge Childs concluded the proceeding with the standard judicial benediction: "May God have mercy upon you." No objection was made to the sentence. Kemmler's trial lawyer, Charles Hatch, was not even present and it appeared "as if Kemmler had been deserted by his lawyers."[109]

Kemmler was shuffled out of the courtroom and taken to the train station. When his party arrived in Rochester at about 7:00 p.m., Warden Durston obtained an order from the superintendent of the New York City railroad to make an unscheduled stop at Port Byron. There Kemmler and his entourage boarded a horse-drawn carriage and continued to Auburn where they entered the prison through the north gate. No one was aware that Kemmler had quietly returned. When the long-overdue Central train arrived at Auburn, a large crowd swarmed the main prison entrance, hoping to sneak a glance at the man about whom so much fuss and frenzy had been made. But after the train pulled out of the station, the crowd soon realized that they had been duped. Kemmler would not be seen in public again.[110]

On August 6, 1890, William Kemmler became the first man to die in the electric chair.

CHAPTER EIGHT

Kemmler's Legacy

The Search for a Humane
Method of Execution

IGHTEEN months after Kemmler's execution, the gag statute was stricken from the electrical execution law. In an overwhelming vote of 105 to 3, the New York Assembly supported a reversal of the press exclusion clause and the state senate followed suit. The new governor, Roswell Flower, formally repealed it on February 4, 1892, when he signed into law a bill striking the ban on press coverage from the Code of Criminal Procedure (Section 507). The revised statute did not require the presence of any specific number of reporters. That decision was left to the warden, who could choose to include members of the press among the twelve "reputable citizens" he was entitled to invite to an execution.[1]

Reporters had been barred from the five electrocutions that occurred between Kemmler's execution and the repeal of the gag statute. This did not, however, prevent public discourse over the humaneness of electrocution. Prison officials, medical doctors, and scientists all insisted that the condemned men died a quick and painless death; most reports in the popular press claimed otherwise. In an editorial the *New York Times*, a staunch supporter of electrocution, questioned the veracity of reports that described unspeakable horrors visited upon the condemned. The editorial went so far as to suggest that the Westinghouse Electrical Company, which was still working to overturn the Electrical Execution Act, deliberately disseminated false reports of executions for the purpose of labeling them cruel and unusual punishment.[2] Most other newspapers greeted the lifting of the press ban with absolute delight, boasting that it would reveal what happened in the state's death chambers.

Charles McElvaine, a nineteen-year-old "young ruffian" from the back alleys of New York City, was the first person executed after the new legislation went into effect. On August 21, 1889, just two days after he was married, McElvaine and two of his "offscourings" broke into a Brooklyn grocery store in the middle of the night. When proprietor Christian Lica woke to discover them, McElvaine repeatedly drove a dagger into the grocer's chest, killing him. A policeman heard Mrs. Lica's cry for help and arrested McElvaine before he could turn his murderous rage on her. McElvaine was tried, found guilty, and sentenced to death.[3]

After two years of appeals, a new trial, and a second conviction, McElvaine was again sentenced to death by electricity. His execution was to take place at Sing Sing Prison, where a separate death house had been constructed. Criminals from the New York City area were sent "up the river"—thirty-three miles—to Sing Sing on the east bank of the Hudson.[4] Warden W. R. Brown (no relation to Harold Brown) invited eight reporters as witnesses, but he would not allow curious members of the public to attend. Armed guards were stationed outside the warden's office with orders to turn back anyone who attempted to enter without an invitation. Arthur Kennelly, Edison's chief engineer and confidant, was among the invited guests, and it would soon become clear why he had been chosen to attend.[5]

The official witnesses stood around Warden Brown's office for more than an hour talking about every conceivable topic except the one that was uppermost in their minds. When the warden finally arrived, the men were escorted through a long winding passage to a small brick building that housed the electric chair. Once inside, the warden warned the men that he would not tolerate any disruption or interference. If he wanted advice from anyone, he would ask for it. Otherwise, he would remove anybody who broke the silence.[6]

The witnesses instantly noticed that the chair was different from those used in previous electrocutions. Instead of broad, flat armrests designed to strap the condemned's forearms to the chair, the new chair had high, sloping armrests that could be used to strap the condemned's underarms to the chair. This design allowed the wrists to be fastened just beneath the level of the seat, permitting the hands to dangle freely in a jar of saline solution. A wire entered one of the jars from below while the other jar was entered from above. The familiar metal scullcap remained suspended from the armature above, although it was not expected to be needed. Beneath the chair lay a rubber mat.[7]

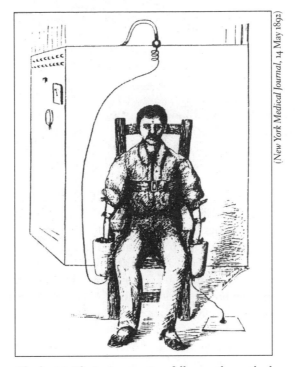

Charles McElvaine's execution, following the method
suggested by Edison

Dr. Carlos F. MacDonald, who had been chief medical counsel at
Kemmler's execution and was now in charge of this execution, told the
assembled witnesses that the electrical contact "will be made through the
hands." Thomas Edison had suggested that to ensure a "perfect contact,"
the prisoner's hands should be immersed in a saline solution. Since
Kemmler's execution, Edison had maintained that the blood in the hands
and arms was a good conductor of electricity, whereas the skull and bones
were poor conductors. In addition, the hair on the head, as a nonconductor, offered resistance to an electrical current.[8] Since this execution was to
be the first officially open to reporters, Warden Brown wanted to be certain nothing went wrong. Just in case, he had the skull and leg electrodes
at the ready.[9]

Without warning, Charles McElvaine entered the death chamber,
accompanied by two Irish Catholic priests. The assembled witnesses let

out a collective sigh. Struggling to maintain his composure, McElvaine walked unassisted the few feet from the doorway to the chair. He held a brass crucifix in both hands, and, staring at it, he mumbled the following prayer: "O Jesus, help me. O Lord. I am sorry that I have offended Thee. O Almighty God, I despise my sins. O Christ, have mercy. Help me, O Lord."[10]

As the guards strapped him to the chair, McElvaine had a look of "abject terror" on his face. He kissed the crucifix one final time. As his right hand was plunged into the jar, Father Creeden removed the crucifix from McElvaine's left hand. In a "frenzy of agony" McElvaine shouted, "Let her go!"[11]

The warden gave the signal—a dropped handkerchief. The current was turned on and McElvaine's body snapped rigid, straining against the leather straps as the muscles contracted wildly. His chest arched out and the strap that held it creaked from the strain. His legs pushed at the straps. One of them drew up sideways. McElvaine's facial muscles became grotesquely contorted, his lips purple, and a froth spilled out of his mouth. The witnesses were horrified. Most feared that the straps would not hold him, that McElvaine's body might vault forward, and his bodily fluid might soil them. The current was kept on for fifty seconds.[12]

After the signal was given to shut off the current, an attending physician reached down into one of the jars to feel for a pulse. At this moment, McElvaine expelled a "great quantity of sputum," followed by a long, protracted groan, "deep and sonorous." Dr. MacDonald panicked and hastily ordered the head and leg electrodes connected. It took almost a minute for this to happen. When the current was turned on again, McElvaine's body convulsed uncontrollably. The witnesses heard a loud hissing sound coming from McElvaine's leg. Some said they smelled human flesh burning; others claimed that it was just steam coming from the scorched sponges. After thirty-six seconds, the current was shut off. Chaos reigned, but McElvaine was dead, undoubtedly.[13]

Dr. Carlos MacDonald stepped forward and declared: "Experience seems to teach us that the contact of the current at the frontal bone and the leg is better than at any other point, and the test with the hands today seems to prove it." While MacDonald was talking, Dr. S. B. Ward examined McElvaine's chest for a heartbeat. When none was found, Dr. MacDonald announced that McElvaine died instantaneously and without pain. One of the attending physicians objected, saying that after the current was shut off the first time, McElvaine remained very much alive.[14]

At his trial, McElvaine's lawyers had alleged that while still a child an ordinary bean became lodged in his ear, causing him to go insane later in life. The autopsy revealed McElvaine to be in good health. His brain showed no sign of imbecility or insanity, nor was any bean found. Professor L. H. Laudy of the Columbia College School of Mines, who had been present at Kemmler's execution, confidently stated that McElvaine's execution was "vastly more satisfactory" than Kemmler's. Kemmler might have been resuscitated after the initial contact, but McElvaine was impossible to resuscitate. Something had been learned between the first and seventh execution.[15]

The official record supported Professor Laudy's contention. It showed the first contact to have lasted for fifty seconds and to have delivered 2 to 3.1 amperes at 1,600 volts. The second contact lasted for thirty-six seconds. It delivered 1,500 volts at a steady 7 amps. Since it is the amperage in conjunction with the voltage that kills, not the voltage alone, the original method of applying the electrodes to the head and leg was deemed superior.[16]

According to Arthur Greaves, the *New York Times* reporter, Mr. Edison was "entirely mistaken" in recommending that the electrodes be applied to the hands and wrists. The record showed that the resistance through the hands was twice as much as through the skull and leg. In an article subtitled "Edison's Idea Failed," Dr. MacDonald was quoted as saying, "Edison probably reasoned all right from his standpoint as an electrician, but all wrong from the standpoint of a physician."[17] This was a mischievous paraphrase of Edison's comments after the Kemmler execution when he complained that the physicians reasoned properly from the standpoint of medicine, but they had no practical knowledge of electricity. It could only have been meant as an insult.

Thomas Edison would not accept defeat. He claimed that his method of executing condemned criminals was a "distinct and pronounced success."[18] He told a reporter that McElvaine did not feel any pain because he became instantly unconscious. As evidence, he offered the fact that the current was kept on for fifty seconds, the longest continuous current of any of the executions. In addition, Edison said that he "never claimed that death would occur more readily, or even as readily, but only that his method was more practical to apply and did not result in excessive burning of the flesh." Arthur Kennelly, who had been sent to Sing Sing to represent Edison's interest, confirmed his boss's interpretation of events, adding that he did not observe anything that would indicate that the condemned man suffered.[19]

Most of the witnesses seemed to agree that McElvaine suffered no pain. Dr. T. S. Robertson, however, insisted that the condemned man was not dead after the first surge of electricity entered his body. A reluctant Dr. S. B. Ward of Albany admitted that when he placed his hand in the jar of saline solution after the current was first turned off, McElvaine had a pulse. And Dr. Robertson told the *Times* reporter that he had seen men "hanged, guillotined, and shot, and he considered the electrical execution the most revolting." He did not consider it "brutal" as much as "brutalizing." "One of its worse features . . . is the effect it must have on the deputies of the prison, . . . the physicians and electricians" who carry out the sentence. "I cannot see," Dr. Robertson continued, "how anybody can regard it as a humane method, or even as a decent one."[20]

Assemblyman Myer J. Stein, who had sponsored the bill that lifted the restrictions on the press, said that after witnessing McElvaine's execution, he never wanted to see another. "It was the most horrible, the most sickening sight I ever witnessed . . . and . . . I feel bound by a sense of duty to humanity" to work to repeal the law. "If the members of the legislature," he continued, "would witness one of its operations they would hold the same views as I do."[21] Before leaving Sing Sing, Stein told reporters that as soon as he returned to Albany he would introduce a bill to replace the electric chair with the rope.

THAT NEVER HAPPENED. Stein's bill failed to pass the assembly. Instead, New York continued to embrace electrocution, and one state after another adopted it as the preferred method of execution: first, Ohio in 1896, then Massachusetts two years later in 1898, followed by New Jersey in 1907, Virginia in 1908, and North Carolina in 1909, Kentucky in 1910, South Carolina in 1912, Arkansas, Indiana, Pennsylvania, and Nebraska in 1913.[22] Despite periodic botches and the occasional gruesome and ghastly execution, the general sentiment developed that electrocution was "less painful and more humane than hanging."[23] Thereafter, states gradually adopted electrocution until 1949 when West Virginia became the twenty-sixth and final state to switch from hanging to the electric chair.[24]

THE ELECTRIC CHAIR did not change much in the next century. It remains the only electrical appliance that has not undergone major modification since its invention more than one hundred years ago.

Indeed, Edward F. Davis, the electrician who wired the original Kemmler chair, still holds the only U.S. patent on the device.[25] Imagine browning bread in a nineteenth-century toaster, or doing the laundry in a washing machine invented in the 1890s. Similarly, the various state criminal statutes stipulating how an electrocution is to be carried out do not differ significantly from the original New York statute. Each statute basically says that the execution should be accomplished by passing through the body of the condemned a current of electricity of sufficient intensity to cause death.[26] After Kemmler's execution the electrode at the base of the spine was moved to the calf of the leg. This provided a more accessible and broader area of thin skin to ensure ample contact. The other electrode is still attached to the top of the head, and, of course, alternating current endures as the executioner's current.

Over the years, however, both the voltage and the amperage employed have been increased: voltage has been increased about 25 percent, whereas amperage has more than doubled. William Kemmler, for example, was slated to receive 1,700 volts. Since no amperage was specified and none measured, we do not know the amount of amperage he received. We do know that Charles McElvaine, the seventh man electrocuted, received 1,600 volts at 2 to 3.1 amperes for fifty seconds and then 1,500 volts at a steady 7 amperes for thirty-six seconds.[27] In the twentieth century, most prisoners have been electrocuted using an alternating current of 2,000 to 2,200 volts at 7 to 12 amperes, for two 1-minute intervals. In New York's last electrocution in 1963, 2,000 volts were applied to the condemned for three seconds, then reduced to 500 volts for fifty-seven seconds. The voltage was again raised to 2,000, lowered for a stretch, and then brought back to 2,000 volts for the remainder of the second minute.[28]

IT IS still not entirely clear how electricity kills. The animal experiments conducted at Edison's West Orange laboratory prior to Kemmler's execution were designed to determine the amount of current necessary to kill briskly without excessive burning of the body—not how electricity actually destroys life. Indeed, the evidentiary hearing ordered by the Cayuga County Court to help determine whether execution by electricity was cruel and unusual punishment focused almost exclusively on the difficulty of measuring the resistance of the average human body to an electrical current. The operating assumption seemed to be that if electrocution was quick and painless, it was not cruel and unusual.

As a result, little attention was paid to the mechanism by which death occurred. Edison himself believed that electricity did not directly destroy the body's vital organs, but instead excited the nerves and muscles to a fatal pitch.[29] Harold Brown thought that electricity killed by causing violent vibrations in body fluids, thus destroying vital organs.[30] Dr. Fell, who conducted experiments on dogs, maintained that the heart's rhythmical movement was at once immediately interrupted by an electrical current. In his presidential address to the American Society of Microscopists— delivered just seven days after Kemmler's execution—he described how under the influence of chloroform and artificial or forced respiration, he cut open the chest walls of several canines to observe the destructive action of electricity on the heart and lungs. Dr. Fell reported that "the heart was beating as in life, but the instant the circuit was made, it ceased its action and became a mere mass of quivering flesh." Since the "ordinary conditions of dying" were absent, the doctor concluded that death by electricity was instantaneous.[31] We now know that this is not the case, at least during an ordinary heart attack. Death is neither sudden nor certain, and a loss of consciousness does not necessarily occur. An ordinary defibrillator—a medical device that administers a counter electrical shock— can be used to stop ventricular fibrillation and restore the heart's normal rhythmical beat.

Recent research indicates that Dr. Fell was the closest to the truth. To be certain, cardiac arrest is the most common cause of death from accidental electrical shocks. Nonetheless, the high voltage used in legally sanctioned electrocutions can also cause death through destruction of the central nervous system.[32] When properly administered, death is certain. The only question that remains is whether death is instantaneous. This is the critical question, however, because experts seem to agree that if death is sudden, then it must also be painless. If death is not immediate, or if it is protracted, as in the case of a botched execution, then electrocution must be excruciatingly painful—like being roasted from the inside.

Dr. Carlos MacDonald, who performed autopsies on the first seven men electrocuted, believed that electricity killed in two ways: by destroying both the central nervous system and the heart. With the attachment of electrodes to the head and leg, the brain and heart are connected in a closed electrical circuit; thus the body's two most essential organs are attacked simultaneously. Dr. Fell explained that the "arrest of the heart's action can be as readily effected by destroying or paralyzing the brain . . . as by attacking the heart itself."[33]

As might be expected, combatants in the debate over the desirability and propriety of capital punishment itself tend to come down on opposite sides of this issue. The general argument of those who claim death is instantaneous and therefore painless is that electrical currents travel much faster than nerve impulses. In a properly conducted electrocution, it takes only 4.16 milliseconds for the condemned to lose consciousness. This is twenty-four times faster than the central nervous system can record pain. To proponents, the conclusion is inescapable: the condemned is dead before he can feel any pain or discomfort.[34]

In legal affidavits presented to courts examining the constitutionality of electrocution, opponents of electrocution have argued that it is far from proven that unconsciousness occurs instantaneously. Harold Hillman, director of the Unity Laboratory in Applied Neurology at the University of Surrey, England, believes that the condemned suffers "severe and prolonged pain." Although "a prisoner being electrocuted is paralyzed and asphyxiated," he is "almost certainly . . . fully conscious and sentient." The condemned's inability to move or failure to display outward signs of pain and agony does not mean that he is spared grave suffering.[35]

Other experts caution that during an electrocution the current might not penetrate the brain completely, since the human skull acts as an insulator, creating more resistance than the skin. This circumstance can be compounded if, due to excessive perspiration caused by extreme nervousness, the skin is dank and clammy. For these reasons and others, Dr. Orrin Devinsky, professor of neurology at the New York University Medical Center, believes that the condemned is likely to experience intense pain, be emotionally traumatized, and lose all perception of time, so that his death appears to last a lifetime. In any event, Dr. Devinsky surmises that no study can definitively establish whether electrocution causes sudden brain death, or even the amount of suffering a person might endure—no one lives to tell us.[36]

THE CONTROVERSY surrounding the execution of William Kemmler did not damage George Westinghouse's business, and alternating current continued to gain ground on Edison's direct current. Within a year, AC had captured more than 50 percent of the lighting market.[37] The Chicago World's Fair of 1893, intended to commemorate the four hundredth anniversary of Columbus's voyage to America, became the next battleground for the two most powerful electric companies.[38] Originally, Westinghouse was not planning to bid on the job of illuminating the Columbian exposition. He knew that much of the exposition would be a celebration of the work of his chief rival, Thomas Edison.[39] And, besides, the Edison Company, now reconstituted as Edison General Electric, held the patent rights on the only available incandescent lamp, and they would undoubtedly refuse to sell them to Westinghouse.[40] To be sure, the Westinghouse Company was challenging these rights in federal court, but they fully anticipated a decision confirming Edison's proprietary rights before the opening of the fair.[41]

It was commonly thought that only the Westinghouse and Edison companies were large enough to handle the illumination of the buildings and grounds of the world's fair. The Edison group, believing it had the field to itself, put in sealed bids ranging from $13.98 to $18.51 per lamp. However, a small Chicago company, the South Side Machine and Metal Works, put in a bid of only $5.49 per lamp. This was a difference of between $1,722,170.40 and $510,789.60 for the 93,040 lamps required.[42] After the directors of the Columbian exposition were assured by the diminutive Chicago enterprise that the Westinghouse Company would manufacture, install, and maintain the system, a second round of bids was initiated. This time the Edison group lowered its bid by two thirds, to $5.95 per lamp, but the Westinghouse Company's new bid of $5.25 was the lowest, and they were awarded the contract. At this price, Westinghouse could not expect to turn a profit, but he apparently felt that the worldwide publicity generated by the exposition and the prestige it would bring to his company made for a sound investment.[43]

The Chicago World's Fair proved an enormous success for Westinghouse and his system of alternating current. The final victory, however, came just weeks before the fair closed. Following a spirited competition with Edison General Electric, Westinghouse was awarded the contract for building a hydroelectric power plant at Niagara Falls.[44] At first, the International Niagara Commission, the group charged with making the decision, expressed a strong preference for direct current. And with Thomas Edison hired as a consultant, it appeared that the Edison companies would be awarded the contract. But after detailed study and much discussion, the commission came to the conclusion that alternating current offered the best hope for the future of the electric power industry.[45]

On June 1, 1892, Edison General Electric merged with Thomson-Houston Company to form General Electric. After Westinghouse, the Thomson-Houston Company was the dominant player in AC power and equipment, surely the major reason it was attractive to the Edison people. General Electric then expanded aggressively into the AC business. Charles Coffin, the chief executive who, as treasurer of the Thomson-Houston Company, had surreptitiously helped Harold Brown obtain the Westinghouse generators for the Kemmler execution, became General Electric's first president. The man behind the merger, however, was J. Pierpont Morgan, the financial baron, who wanted to consolidate the electrical power industry and stifle competition as he had done with the railroads, oil, coal, and steel industries.[46]

Edison was against the merger, but by this time he had been muscled to the sidelines. He felt that Edison General Electric was still the dominant electric company in America, "not only in prestige but in profit," and that "no competition means no invention."[47] Embittered that his name, which had become synonymous with the electrical power industry, had been dropped from the new company, Edison soon resigned from the board of directors. Sitting in his parlor several weeks after the merger, he told a friend that he was finished with electric lighting and power. "I am going to do something now so different and so much bigger than anything I've ever done before that people will forget my name ever was connected with anything electrical."[48]

Gradually, direct-current power plants started switching over to alternating current. By 1894 a rotary converter was in operation, making possible the linking of long-distance, high-voltage AC transmission lines with DC distribution networks.[49] And, in 1896, after some difficult negotiations, General Electric and Westinghouse signed a deal granting each other manufacturing licenses to all patents except those applicable to electric lighting.[50] With this agreement, DC began to fade away, although isolated power plants still existed until after World War II. Today, American households are on a standard universal system of alternating current at 110 volts and 60 hertz.[51] However, most household appliances (for example, telephone answering machines, radios, and portable computers, which otherwise run on direct current from a battery)—and all electronics—actually run on direct current. Inside, they contain a transformer that converts alternating current to direct current for use at various voltages.

Meanwhile, Edison turned his attention to other matters. In 1893 he developed the first motion-picture studio, a tar-papered shack nicknamed "the black Maria." A year later, he and an employee, William K. L. Dickson, invented a kinetograph, the first practical motion-picture camera, and the kinetoscope, the first movie projector. Another of Edison's accomplishments after the battle of the currents was the perfection of the alkaline storage battery, originally developed as a means of powering his phonograph in homes without electricity. By 1909 Edison was a main supplier of submarine batteries and had begun work on a new storage battery that produced electricity through a chemical reaction. Believing it would replace the lead battery, a heavy, difficult-to-recharge battery using the highly corrosive sulfuric acid, Edison formed a company for the manufacture of electric motor cars.[52] Unfortunately, he never was able to perfect his nickel-iron-alkaline storage battery. A few years later, he designed a car battery for the Ford Model T, the first motor vehicle with an automatic ignition.

During World War I, Edison conducted research for the U.S. Navy on forty-eight different projects, from modified torpedoes equipped with turbines to spray-free scopes for submarine periscopes. He also worked on designing a poured Portland cement house "to stop the ruthless devastation of pine forests" and "place homes within the reach of thousands."[53] His biggest failure was his futile attempt throughout the mid-1880s and 1890s to devise a magnetic ore-separator, a device designed to harvest platinum from sand rich in iron.

Edison did, of course, prevail in the area of reputation: his hand in the Kemmler affair was eventually forgotten, overshadowed by his enormous achievements. He is remembered today as one of the most significant, and beloved, figures in American history. Cosseted by honors, the great man did eventually admit, albeit privately, that he was wrong about the future of alternating current. In October 1929, on the fiftieth anniversary of Edison's invention of the lightbulb, the nation paid tribute to him on what was dubbed Light's Golden Jubilee.[54] The celebration took place at the Edison Institute—the permanent birthplace of light—an eight-acre museum built by his friend, Henry Ford, in Dearborn, Michigan. The automobile manufacturer had moved Edison's Menlo Park laboratory and the railroad station—where, as a young boy, Edison was tossed out after setting fire to the train while conducting one of his many boxcar experiments. Edison died on October 18, 1931, and was buried in West Orange, New Jersey. He was eighty-four years of age. In 1962 his West Orange laboratories and his twenty-three-room mansion, Glenmont, were designated a national historic site. The laboratory, machine shop, and patent office are as he left them, including his library, scientific papers, and professional notebooks, along with early models of his many inventions.[55]

· · ·

AFTER NIAGARA FALLS, George Westinghouse went on to improve alternating-current systems and to refine the steam turbine. At the height of his career, he employed more than fifty thousand people. Despite his mercantile acumen, Westinghouse had taken on too much debt, and the financial panic of 1907 caused several Westinghouse companies to fail. Their reorganization and eventual solvency was based upon a plan devised by Westinghouse, but the actual control of the Westinghouse Electric and Manufacturing Company passed out of his hands— although he stayed on as its titular head until 1911, when he resigned. The following year, the American Institute of Electrical Engineers bestowed on him its highest honor—the Edison Medal.

Afterward, Westinghouse continued to invent and improve upon various devices. He is remembered chiefly for the introduction and dissemination of the alternating-current system of electric power transmission, as well as for his invention of the air brake, the gas meter, and a method of transporting natural gas safely into homes and businesses. One of the most successful entrepreneurs of his day, he was a resourceful manager and intelligent businessman, in addition to being a first-rate engineer and inventor. Although nowhere near as prolific as his rival Edison, Westinghouse nonetheless holds patents on 361 inventions.[56]

Throughout most of the twentieth century General Electric and Westinghouse Electric Company continued to compete head-to-head in the marketplace. Today, General Electric's businesses include aircraft engines, appliances, capital services, electrical distribution and control, information services, lighting, medical services, industrial control-systems, plastics, power and transportation systems, and the NBC television network. In 2002, General Electric made three of *Fortune* magazine's lists of notable companies. It currently ranks as the eighth largest company in the world, and first in the world's most-admired companies. While General Electric pursued a strategy of diversification during the twentieth century, the relative stability of its leadership has allowed the company to flourish for more than a century.[57]

The Westinghouse Company did not fare as well. For most of the twentieth century Westinghouse prospered. It was responsible for the development of the radar first used at Pearl Harbor, the first U.S.-designed jet engine, and the first U.S. nuclear power plant. Then, in the late 1980s, following a string of failed CEOs and diversification strategies, Westinghouse faltered. Beginning in 1988, Westinghouse began a series of divestitures that included the sale of electrical supply, defense electronics, and land development businesses. In 1997, Westinghouse was forced to put its remaining industrial divisions up for sale, consolidating the commercial-nuclear-power business and media operations, including CBS radio and television networks, under the CBS corporation. In 1999, British Nuclear Fuels acquired the nuclear-power business from CBS, operating it currently under the name of Westinghouse Electric Company. In a little more than ten years, the Westinghouse Company went from a major corporation to a modest, wholly owned subsidiary.[58]

HAROLD BROWN went on to have a highly successful career both as an inventor and as an independent consulting engineer in New York City. Throughout the years, he served as a consulting and erecting engineer to several railroad and manufacturing companies and to some of the nation's largest cities. In 1898 he received the John Scott medal from the Franklin Institute in Philadelphia for the development of "plastic rail-bond," a compound that allowed the train rails to carry heavy currents while maintaining the necessary flexibility. For his "special alloys" for electrical contacts, he received the highest honors at the Paris Exposition in 1900. Thirteen years later he invented an "apparatus for lining railroad tunnels with concrete." He also invented an early hydraulic brake for automobiles and a waterproof concrete for boat and ship construction—designed to withstand repeated torpedo attacks—for the U.S. Navy. His work with concrete led to his employment in home construction by the British health ministry from 1920 to 1923. Harold Pitney Brown holds fifty-seven patents.[59]

For almost four decades after Kemmler's execution, as far as can be determined, Edison and Brown had no contact with each other, although the two men were very much alike—self-taught, independent-minded, old-fashioned electricians. And both were not only inventors of electrical devices, but were also in the cement/concrete business. Like his onetime mentor, though, Brown never stopped dreaming of capturing the public's imagination by coming up with a truly life-changing or earth-shattering invention. On February 10, 1927, thirty-seven years after the two men last parted company, Brown wrote Edison a letter[60] hoping to renew their professional relationship. After congratulating Edison on his eightieth birthday (when Brown was sixty-nine years old himself) and thanking him for his "great kindness . . . when I was fighting for my life in the contest with the 'safe and harmless' Westinghouse alternating current," Brown said that he had at last found a way to express his gratitude.[61]

In his letter, Brown touted the virtues of his latest invention — "an electro-chemical method of removing fatigue poisons from the system which restores capacity for physical and mental work."[62] Recalling their prior work together, Brown wrote that he would be glad to offer his invention to Edison so that it could be "chemically tested" in Edison's laboratory. "[A]s in the case of the 'Electrical Duel' with Mr. Westinghouse, I am willing to submit my self [sic] for the demonstration." Edison, to the surprise of no one, except perhaps Brown, was not interested. He scribbled a note on the top of Brown's letter, instructing his assistant, William H. Meadowcraft, to "get me out, don't believe in his thing."[63]

A month later, Meadowcraft wrote Brown saying that he was sorry for the delay in getting back to him, but his letter had been misplaced and Edison was presently in Florida on vacation. "From experience, you know how things go around there, and I am quite sure you will understand."[64] Two months passed before Meadowcraft wrote Brown informing him that Edison was much too busy with "special experiments" to give his invention the attention it deserved. Meadowcraft enclosed an autographed picture of Edison.[65]

WILLIAM KEMMLER became a minor but memorable character in the dual history of capital punishment and the battle of the currents. He remains the only one not diminished by his role in the ordeal. Indeed, he seemed elevated by it. During his eighteen months on death row, he had learned to read and sign his name, he had studied the Bible under the tutelage of the warden's wife, he had given up tobacco, regained his health, and made peace with his Maker. Kemmler faced his execution "like a man," calm and collected, dignified and courageous, with the presence of mind to help the warden tighten the leather straps to his body and secure the electrode to his head. And, most of all, he reassured the witnesses — men of science and medicine — that he would follow the execution etiquette and not create a scene. He believed he was going to a better place.

A century after his death, Kemmler's great-grandnephew uncovered the long-hidden skeleton in his family's closet and decided that his ancestor's prison redemption, exceptional grace under pressure, and touching faith in science and humanity outweighed the terrible shame of his crime. Kemmler's family came to be proud of the way he died, if not the way he lived. Today, Kemmler's family has become middle-class, with Kemmler's great-grandnephew the first to graduate from college. They still live in Philadelphia, but like many successful families, they have moved to the suburbs.

No trace of little Ella could be found.

FIFTY-FOUR MEN and one woman died in the electric chair at Auburn. There were chairs in Auburn, Clinton, and Sing Sing until 1914. After that, all executions took place at Sing Sing until 1963, when New York State last electrocuted a convicted criminal.[66] In 1971, the chair, along with its electrical equipment and witnesses' pews, was moved to the Green Haven Correctional facility, where it has never been used.[67] In all, 686 men and 9 women died in New York State's three electric chairs. When capital punishment was reenacted in 1995, lethal injection was substituted for electrocution. The new death chamber is located in the Clinton Correctional Facility at Dannemora. Over the years several people have claimed they had the original chair, but the Kemmler chair was destroyed by fire in a prison riot at Auburn in 1929.[68]

IN THE TWENTIETH CENTURY more than forty-three hundred people were legally executed by electricity, exceeding all other methods combined. Even though the U.S. Supreme Court never decided whether electrocution was cruel and unusual punishment, courts have continued to rely on *In re Kemmler* for the proposition that electrocution is permissible under the Eighth Amendment of the U.S. Constitution. Moreover, *Kemmler* has been relied on to support the constitutionality of all other methods of execution employed throughout our nation's history, from shooting to hanging to lethal gas to lethal injection. This reliance is particularly misplaced when applied to hanging, the method the New York legislature replaced with electrocution.[69] To date, the U.S. Supreme Court has never reviewed evidence regarding the constitutionality of any method of execution.

Today, *Kemmler* generally represents the proposition that a punishment may be constitutional even if it is unusual, as long as it has a humanitarian purpose and effect.[70] But the most significant and still legally sound contribution of *In re Kemmler* to Eighth Amendment jurisprudence was the standard of cruelty it proposed—a standard that would be cited repeatedly throughout the next century. "Punishments are cruel when they involve torture or a lingering death . . . [when] there is something inhumane and barbarous, something more than the mere extinguishment of life."[71]

In 1972, in *Furman v. Georgia*, the U.S. Supreme Court held that the death penalty as administered was unconstitutional because capital cases resulted in arbitrary and capricious sentencing. The Court held that Georgia's death-penalty statute was "cruel and unusual" and a violation of the Eighth Amendment because it gave the trial jury unguided discretion to impose a sentence of death. Since Georgia's law was similar to those of other states, the Court effectively voided all forty death-penalty statutes.[72] Almost immediately, states began to rewrite their death-penalty laws to eliminate the arbitrariness cited in *Furman*. A de facto moratorium on executions, which began in 1967 with legal challenges to capital punishment, lasted until 1977 when Gary Gilmore was executed in Utah by firing squad.

The resumption of executions—and the uneasiness that accompanied them—provided the impetus for states to begin replacing the electric chair with lethal injection.[73] Today, ten states still authorize electrocution, but only Nebraska relies on it exclusively. Of the other nine states, four (Alabama, Florida, South Carolina, and Virginia) allow prisoners to choose between electrocution and lethal injection. Three states (Arkansas, Kentucky, and Tennessee) also allow prisoners to choose between electrocution and lethal injection, but only if their crimes were committed before a particular date. And two states (Illinois and Oklahoma) authorize electrocution, should lethal injection ever be held unconstitutional. There can be little doubt, however, that the electric chair's days are numbered, and we may well have seen the last of its use.[74]

THIS BRINGS US TO THE QUESTION that has impelled the writing of this book. Can an execution ever be considered humane? The answer undoubtedly depends on what we consider to be the meaning of the word *humane*. If *humane* means behaving toward fellow human beings in a way that does not exhibit gratuitous callousness or incivility, or if *humane* is used merely to describe methods of punishment that inflict the least amount of pain, then it is certainly possible to have a humane execution, or at least one that is not bloody and brutal. But if *humane* means acting in a way that benefits human beings, behavior that respects the sovereignty of the individual while maximizing his human potential, then no execution can be considered humane.

The difficulty in answering the question lies, perhaps, in the unacknowledged truth that we are as concerned with our own suffering as with that of the condemned. The entire history of the death penalty has been driven not so much by humanitarian concerns as by the necessity to avoid offending the human sensibilities of those in whose name an execution is carried out. This explains our long-standing concern with the aesthetics of executions. The simple fact is that if all one wants to do is kill, execution is easy, if not always pretty. Death by the rope is brutal, gruesome, and often lengthy. Gas can be slow and cause excruciating contortions. A bullet to the head is quicker and probably less painful, but it is messy. We know what electricity can do to a human body. It is our anguish over appearances that suggests our profound ambivalence regarding execution, which is, finally and inescapably, the taking of a human life.

At its core, the debate about capital punishment is not about methods of execution—the methods are nothing more than a distraction. The concept of humane punishment has come to mean punishment that conceals its violent nature. If the true violence is killing, then all methods of executions are equally violent. Today's executions are perceived as more humane; they are less passionate, more impersonal—relying upon a highly rationalized machinery of death that emphasizes the efficiency of killing rather than the morality of the law. Capital punishment only appears to be getting more humane, for we no longer treat the condemned as a real human being. The extensive process of dehumanization that occurs on death row helps to assure that the condemned is already dead long before he enters the death chamber.

In surveying the entire history of capital punishment, it is most obvious that every so-called advance in the means of execution has been in response to a rise in opposition to the death penalty. The shift from public to private executions, from hanging to electrocution, from the chair to the hospital gurney—each has represented an attempt to defang death-penalty opponents by making the process appear more humane. And yet the unspoken truth is that the humanity we are protecting is our own—humankind's natural revulsion at the prospect of taking even the most evil life. By sanitizing the process with advances in science and medicine—

with the idea of progress—we have managed to turn a public ceremony, designed to venerate the power of the law and the primacy of the soul, into a private procedure that happens offstage as part of a consoling ritual known as justice. It is for this reason that methods calculated to meet the "cruel and unusual" standard so often end in appalling savagery and why the legacy of the electric chair is not "quick and painless" but the very opposite. William Kemmler's hideously botched execution arose not from scientific incompetence or business chicanery but from the irreconcilable division within our own soul.

Perhaps our objections to killing another human being could be overcome if we had a clear consensus on the deserved nature of punishment. But we are a society distrustful of courts and other public institutions, suspicious of our own values, and deeply committed to the belief in the sovereignty of the individual. In order to impose the ultimate penalty, we need to place full responsibility on the individual for his behavior. The concept of full responsibility, however, is better fitted to an earlier age. We know too much about the workings of the human mind, the vagaries of human nature, the political, economic, and cultural components to which crime and punishment are intricately related. Our understanding of unlawful behavior no longer rests solely on a nineteenth-century free-will view of human motivation. If it did, executions would still be public and justice would not have to be served in private.

The story of the invention of the electric chair reveals how science, technology, and progress first became linked in the application of capital punishment. The involvement of Thomas Edison and others dramatically illustrates how a moral issue of "humaneness" was transformed into a technological problem of "quick and painless," and eventually into the legal question of "cruel and unusual."

In the end we are left with many profound and disturbing questions—not only about electrocution, but about the scientific and technological nature of our search for a humane method of execution. Someday we may be clever enough to devise a quick, painless, and even reliable method of execution. But let's not delude ourselves: No method of execution can ever be considered humane.

NOTES

INTRODUCTION

1. Richard Moran, "Invitation to an Execution—Death by Needle Isn't Easy," *Los Angeles Times*, 24 March 1985, section 4, p. 5.

CHAPTER ONE "William, It Is Time"

1. "Far Worse Than Hanging," *New York Times*, 7 August 1890, p. 1.

2. Only one of the three commissioners responsible for the electrocution law, Dr. Alfred P. Southwick, attended the execution. As one of the major forces behind the new Electrical Execution Act of 1888, he did not want to miss this moment. Noticeably absent was Elbridge T. Gerry, founder of the Society for the Prevention of Cruelty to Animals and chairman of Governor David Hill's commission investigating alternative methods of execution. The "commodore," as Attorney Gerry was affectionately known, was sailing at the New York Yacht Club, confident in his belief that death by electrocution was undoubtedly the most painless method of execution. The third and final member of the commission, Matthew Hale, an Albany attorney, declined the invitation without explanation. The idea for the commission had originated with former New York state senator Daniel MacMillan of Buffalo, who was in Washington that fateful morning. Senator Saxton, who introduced the electrocution bill to the assembly floor, also failed to attend.

3. "Killed! Kemmler Is Put to Death by the Electric Process," *Buffalo Evening News*, 1st ed., 6 August 1890, p. 1.

4. The following is a list of witnesses to Kemmler's execution: Dr. H. E. Allison, head of the state asylum for the criminally insane in Auburn; Dr. Henry A. Argue of Corning; George G. Bain, a reporter for the United Press; Dr. Louis Balch, secretary of the State Board of Health; Tracy C. Becker of Buffalo, who served as referee in Kemmler's trial; Dr. Michael Conway of Troy; Dr. C. M. Daniels of Buffalo; New York electrician E. F. Davis, sent to Auburn by Harold Brown; electrician Robert Dunlap of New York; Dr. George E. Fell of Buffalo, famous for the resuscitation of electrocuted animals and also a professor of physiology at Niagara University; Dr. Joseph Fowler of Buffalo, who performed the first autopsy on an accidental electrocution victim; Buffalo electrician C. R. Huntley; execution assistant George W. Irish of Albany; Dr. J. M. Jenkins of Auburn; Erie County sheriff Oliver A. Jenkins; New York deputy coroner Dr. William T. Jenkins; Dr. Carlos F. Mac-Donald of New York, president of the State Lunacy Commission; Associated Press reporter Frank W. Mack; Dr. W. J. Nellis of Albany; Dr. George F. Shrady of New York, who supervised most of the autopsy; Dr. T. K. Smith of Auburn; Dr. A. P. Southwick; Dr. E. C. Spitzka, who held the stopwatch, pronounced Kemmler dead after the first shock, and supervised the autopsy of the brain and spinal column; and Under-sheriff Joseph C. Vieling of Erie County. "No More Hope for Him," *Buffalo Evening News*, 5 August 1890, p. 1.

5. "Far Worse Than Hanging."

6. "The Death Warrant: Official Document Which Seals Kemmler's Fate," *Buffalo Commercial Advertiser*, 17 May 1889, p. 1.

7. "The Execution Described," *New York Tribune*, 7 August 1890.

8. "One Fell Swoop," *Buffalo Evening News*, 14 August 1890, p. 1.

9. "Kemmler's Last Hours," *Buffalo Evening News*, 5 August 1890, p. 1.

10. "Kemmler Makes His Will: A Belief That He Will Meet His Fate To-Morrow," *New York Times*, 29 April 1890.

11. "Kemmler: Sentenced to Die by Electricity During the Week of August 4," *Buffalo Evening News*, 12 July 1890, p. 1.

12. "Kemmler Makes His Will."

13. "Kemmler's Song," *Buffalo Evening News*, 26 May 1889.

14. Ibid.

15. "Kemmler's Last Hours."

16. "Hopeful and Hopeless," *Buffalo Evening News*, 6 August 1890, p. 4.

17. "The End Is Near at Hand," *Buffalo Evening News*, 5 August 1890, p. 1.

18. "Believes It Will Work," *Buffalo Evening News*, 4 August 1890, p. 1.

19. "Death Is Drawing Nearer," *New York Times*, 5 August 1890, p. 1.

20. "Last Days of Kemmler," *New York World*, 1 August 1890, p. 1.

21. "Kemmler's Last Days: During Next Week the Murderer Will Be Electrocuted," *Buffalo Evening News*, 1 August 1890.

22. "The Last Days of Kemmler."

23. "He May Be Killed To-Day," *New York Times*, 4 August 1890, p. 1.

24. "History of the Dead Murderer: The Wealthy Corporation that Fought to Save Kemmler from His Doom," *New York Tribune*, 7 August 1890.

25. "Westinghouse and Sherman," *New York Sun*, 9 August 1890, p. 1.

26. "History of the Dead Murderer."

27. *People ex. rel. Kemmler v. Durston*, 7 N.Y.S. 813 (Sup. Ct. 1889), *In re Kemmler*, 136 U.S. 436 (1890).

28. "Kemmler: Sentenced to Die."

29. Ibid.

30. "Tomorrow Night: Kemmler Will Be Executed Sometime After 7 O'Clock in the Evening," *Buffalo Evening News*, 4 August 1890, p. 1.

31. "Kemmler and Fish," *Auburn Daily Advertiser*, 14 July 1890.

32. "Kemmler Will Be First," *Buffalo Evening News*, 17 July 1890, p. 1.

33. "It Might Be Today," *Buffalo Evening News*, 4 August 1890, p. 1.

34. "Kemmler's Last Sunday," *New York Sun*, 4 August 1890.

35. "Kemmler! He May Live Till Thursday, But No One Seems Certain," *Buffalo Evening News*, 4 August 1890, p. 1.

36. "Kemmler Nearing His End," *New York Sun*, 5 August 1890, p. 1.

37. "Kemmler's Last Night," *New York Times*, 6 August 1890, p. 1.

38. "He May Be Killed To-Day."

39. "Kemmler's Last Night."

40. Ibid.

41. "Kemmler's Baptism and Communion," *Buffalo Evening News*, 6 August 1890, p. 1.

42. "Kemmler's Last Night."

43. Ibid.

44. Hence, Kemmler would later hear the witnesses as they descended the stairs.

45. "Kemmler's Last Night."

46. "William Kemmler Dead," *New York Tribune*, 7 August 1890.

47. Ibid.

48. Charles R. Huntley, the Brush Electric Light Company manager, described the electrical components and dynamo in a letter to the editor of the *Electrical World*. He said the engine was "belted to a 2-inch shaft which ran through the dynamo room. On this shaft were a 36-inch pulley on the dynamo, and a 6-inch belt ran from the shaft to the exciter. . . . The wires led out through the window up over the roof on porcelain and glass insulators, down, across and around the dome of the front of the prison; from there down on the wall into the window of what was known as the original room for the electrocution."

49. "It Is Guess Work," *Buffalo Evening News*, 16 August 1890, p. 1.

50. "Killed! Kemmler Is Put to Death by the Electric Process," *Buffalo Evening News*, 2nd and 3rd eds., 6 August 1890, p. 1.

51. "It Is Guess Work."

52. "Far Worse Than Hanging."

53. Ibid.

54. "Killed! Kemmler Is Put to Death," 5th ed.

55. "Description of the Death," *Buffalo Evening News*, 6 August 1890, p. 1.

56. "Killed! Kemmler Is Put to Death," 5th ed.

57. "Was He Tortured?" *New York Press*, 7 August 1890, p. 1.

58. "Description of the Death."

59. "Far Worse Than Hanging."

60. "Killed! Kemmler Is Put to Death," 5th ed.

61. "Far Worse Than Hanging."

62. "Killed! Kemmler Is Put to Death," 2nd and 3rd eds.

63. "A Little More Than 1,000 Volts," *Buffalo Evening News*, 6 August 1890, p. 4.

64. "Far Worse Than Hanging."

65. "Was He Tortured?" Later, both Dr. Fell and Dr. Southwick would agree that Kemmler had displayed great courage and that he did all he could to assist the executioners. Fell, who stood next to Kemmler while he was being strapped into the chair, said that he did not show "even a tremor" and that after the electrodes had been applied to his head, spine, and arms, he never "evinced the slightest fear."

66. "Was He Tortured?"

67. "Irish Held the Lever," *Auburn Daily Advertiser*, 11 August 1890.

68. "Kemmler's Executioner," *Buffalo Evening News*, 7 August 1890, p. 4. There are several conflicting accounts of who actually threw the switch. Warden Durston claimed that a convict in the dynamo room was responsible for it. But in view of Irish's history of interest in executions and public pronouncements, it seems the warden was merely trying to divert attention from the executioner's identity.

69. "Kemmler Continued," *Auburn Daily Advertiser*, 7 August 1890.

70. "A Little More Than 1,000 Volts."

71. "Kemmler Continued."

72. "It Was a Horror," *Buffalo Evening News*, 7 August 1890, p. 1.

73. "Far Worse Than Hanging."

74. "A Little More Than 1,000 Volts."

75. "The Apparatus Used in the Electrical Execution," *Electrical World* 15, no. 18 (3 May 1890): 43.

76. "A Secret of the Kemmler Execution," *Buffalo Evening News*, 11 August 1890, p. 1.

77. "Kemmler Continued."

78. "It Was a Horror."

79. "Dr. Southwick Talks," *Buffalo Evening News*, 6 August 1890, p. 4.

80. "The Autopsy," *Buffalo Evening News*, 6 August 1890, p. 4.
81. Ibid.
82. "The Execution Described."
83. "Execution by Electricity," *The Times* (London), 7 August 1890.
84. "The Execution Described."
85. "Pickled Kemmler: Preparing Segments of His Anatomy for Scientific Observation," *Buffalo Express*, 9 August 1890.
86. "The Autopsy."
87. "Kemmler Nearing His End."
88. "Burial of Kemmler," *Auburn Daily Advertiser*, 8 August 1890.
89. "Kemmler's Body Buried," *Albany Evening Union*, 8 August 1890.
90. "Kemmler's Funeral," *Buffalo Evening News*, 8 August 1890, p. 1.
91. "Burial of Kemmler."
92. "Kemmler Buried in Great Secrecy," *Buffalo Courier*, 9 August 1890.
93. "An Unpretentious Burial: The Murderer's Remains Hurriedly and Secretly Interred," *Buffalo Express*, 9 August 1890.
94. "Buried in Quicklime," *Buffalo Evening News*, 7 August 1890, p. 4.
95. "A Failure: The Electric Torture of William Kemmler," *Buffalo Express*, 8 August 1890.
96. "The Tragedy at Auburn," *New York Sun*, 8 August 1890, p. 1.
97. "Doctors Who Disagree," *Buffalo Evening News*, 3rd ed., 8 August 1890, p. 2.
98. "Liar and Scoundrel," *Buffalo Evening News*, 11 August 1890, p. 1.
99. "Dr. Spitzka Raps Dr. Daniels," *Buffalo Evening News*, 8 August 1890, p. 1.
100. "Liar and Scoundrel."
101. "Kemmler Continued."
102. "Exit Kemmler," *Evening News and Telegraph*, 6 August 1890.
103. "Doctors Who Disagree."
104. "The Execution Described."
105. "Kemmler's Autopsy," *Buffalo Evening News*, 7 August 1890, p. 1.
106. "What the Doctors Think," *New York World*, 8 August 1890, p. 2.
107. Ibid.
108. "Was He Tortured?"
109. "Kemmler's Autopsy."
110. "One Fell Swoop."
111. "Kemmler's Last Hours."
112. "After the First Shock," *Buffalo Evening News*, 8 August 1890, p. 4.
113. "Dr. Fell Didn't Try It," *Buffalo Evening News*, 8 August 1890, p. 1.
114. "Was He Tortured?"
115. "Electrocution by Electricity: Dr. Shrady Does Not Fully Approve of the New Method," *Buffalo Evening News*, 7 August 1890, p. 1.
116. "Doctors Who Disagree."
117. Ibid.
118. "Sheriff Jenkins Is Thankful," *Buffalo Evening News*, 7 August 1890.
119. Ibid.
120. "Aroused by Kemmler's Death: Efforts to Save Other Murderers from a Like Fate," *New York Times*, 9 August 1890, p. 1.
121. Charles R. Huntley, interview, *Buffalo Evening News*, 8 August 1890.
122. "The Point at Issue," *Buffalo Courier*, 8 August 1890.
123. "Negligence: Why the Execution Was a Failure," *Buffalo Express*, 8 August 1890.

124. "It Is Guess Work."

125. "Warden Durston in Albany: He Says the Electrical Mode of Killing Is Far Preferable to Hanging," *New York Sun*, 9 August 1890, p. 1.

126. "Gov. Hill Favors Electrocution," *Buffalo Evening News*, 8 August 1890, p. 1.

127. "Catechising the Governor," *Albany Times*, 8 August 1890.

128. "A Great Day's Work," *Buffalo Evening News*, 6 August 1890.

129. "Newspapers and Electric Executions," *Buffalo Courier*, 8 August 1890.

130. "A Failure: The Electric Torture of William Kemmler."

131. "Burning at the Wire," *New York Press*, 7 August 1890.

132. "Electrocution Repeal Demanded," *Buffalo Evening News*, 7 August 1890, p. 1.

133. Ibid.

134. See Michael Schudson, *Discovering the News: A Social History of American Newspapers* (New York: Basic Books, 1981) for further details about New York newspapers at the turn of the century.

135. "Electrocution Repeal Demanded."

136. "Kemmler's Death Discussed," *Albany Times*, 7 August 1890.

137. Ibid.

138. "Death by Electricity," *New York Tribune*, 7 August 1890.

139. "Repeal the Law," *Buffalo Evening News*, 8 August 1890, p. 2.

140. "Electrocution Repeal Demanded."

141. "The First and the Last," *New York Sun*, 7 August 1890.

142. "Electrocution Repeal Demanded."

143. "Never Mind the Intentions," *Buffalo Evening News*, 8 August 1890, p. 2.

144. Ibid.

145. "A Failure."

146. "Newspapers and Electric Executions."

147. "This Death? No Death!" *Buffalo Express*, 8 August 1890.

148. "One Is Enough," *Buffalo Evening News*, 7 August 1890, p. 1.

149. "Electrocution Repeal Demanded."

150. "The Execution," *Auburn Daily Advertiser*, 7 August 1890.

151. "Kemmler," *Auburn Daily Advertiser*, 12 August 1890.

152. Ibid.

153. "Kemmler's Execution," *Auburn Daily Advertiser*, 8 August 1890.

154. "What Is Going On Abroad?" *Buffalo Evening News*, 7 August 1890, p. 4.

155. "A Failure: The Electric Torture of William Kemmler."

156. "London Opinion: Nearly Every Paper Condemns Kemmler Execution," *Buffalo Courier*, 8 August 1890.

157. "What Is Going On Abroad?"

158. "A Westinghouse View of the Execution," *Electrical World* 16 (16 August 1890): 105.

159. "Aroused by Kemmler's Death," *New York Times*, 9 August 1890, p. 1.

160. Ibid.

161. "The Last of Kemmler," *Buffalo Express*, 9 August 1890.

162. "Westinghouse, Jr., Talks," *Buffalo Courier*, 9 August 1890.

163. "Opinions of Electricians," *New York Tribune*, 7 August 1890, p. 1.

164. "Mr. Edison on the Kemmler Execution," *Electrical World* 16 (16 August 1890): 105.

165. Ibid.

166. "Doctors Who Disagree."

167. "Mr. Edison on the Kemmler Execution."

168. Ibid.

169. The following sources were also consulted in researching and writing this chapter: "Abolish the Death Penalty," letter to the editor, *Auburn Daily Advertiser*, 12 August 1890; "A Protest," letter to the editor, *Auburn Daily Advertiser*, 12 August 1890; "Execution of the Death Penalty," *Frank Leslie's Illustrated Newspaper*, 11 July 1885, p. 330; "Five Murderers Awaiting Execution," *Buffalo Evening News*, 11 August 1890, p. 1; Michael Madow, "Forbidden Spectacle: Executions, the Public and the Press in Nineteenth-Century New York," *Buffalo Law Review* 43, no. 2 (fall 1995), pp. 461–562; "Repeal the Electrocution Law!" *Electrical Review* 18 (15 August 1890): 324; "Snap Judgement," *Frank Leslie's Illustrated Newspaper*, 23 August 1890, p. 43; "The New Death Penalty," *Albany Times*, 8 August 1890.

CHAPTER TWO The Battle of the Currents:
Edison versus Westinghouse, DC versus AC

1. John W. Howell and Henry Schroeder, *History of the Incandescent Lamp* (Schenectady, N.Y.: Maqua, 1927), p. 26.

2. Matthew Josephson, *Edison, A Biography* (New York: McGraw Hill, 1959), pp. 117, 183.

3. George S. Bryan, *Edison: The Man and His Work* (Garden City, N.Y.: Doubleday, 1926), pp. 106–7.

4. Harold C. Passer, *The Electrical Manufacturers 1875–1900* (Cambridge, Mass.: Harvard University Press, 1953), pp. 80, 82.

5. Howell and Schroeder, p. 26.

6. Ibid., p. 32.

7. Ronald W. Clark, *Edison: The Man Who Made the Future* (New York: Putnam's Sons, 1977), p. 90.

8. This standard is still used today.

9. Clark, p. 90.

10. Passer, p. 85.

11. Ibid.

12. Margaret Cheney, *Tesla, Man Out of Time* (Englewood Cliffs, N.J.: Prentice-Hall, 1981), p. 32.

13. Kathleen McAuliffe, "The Undiscovered World of Thomas Edison," *The Atlantic Monthly* (December 1995): 80–93.

14. Josephson, pp. 212–15.

15. Ibid.

16. Clark, p. 97.

17. McAuliffe.

18. Passer, pp. 80–81.

19. Josephson, p. 223.

20. Ibid.

21. Ibid., p. 224.

22. Ibid., p. 225.

23. Ibid., p. 226.

24. Ibid., p. 225.

25. Clarice Stasz, *The Vanderbilt Women* (New York: St. Martin's Press, 1991), p. 86.

26. "M. De Moncel," *New York Times*, 21 January 1880, p. 4.

27. David O. Woodbury, *Beloved Scientist: Elihu Thomson* (New York: McGraw-Hill, 1944), p. 109.

28. At one time in the early 1880s, Edison was said to be working on forty-four different inventions.

29. Josephson, pp. 136–37.

30. Ibid., pp. 254–55.

31. Passer, pp. 90–91.

32. Josephson, p. 261.

33. H. L. Satterlee, J. *Pierpont Morgan* (New York: Macmillan, 1939), p. 156.

34. *Thirty Years of New York, 1882–1912: Being a History of Electrical Development in Manhattan and the Bronx* (New York: Press of the New York Edison Company, 1913), p. 33.

35. Passer, pp. 92, 121–22.

36. Josephson, p. 257.

37. Passer, p. 90.

38. Josephson, p. 257.

39. Ibid., pp. 257, 263.

40. Ibid., pp. 263–64.

41. Thomas P. Hughes, *Thomas Edison, Professional Inventor* (London: H. M. Stationery Office, 1976), p. 33.

42. Josephson, p. 264.

43. From September 4, 1882, to January 2, 1890, the station ran continuously except for three hours. On January 2, 1890, a fire shut down the station. Ten days later it resumed operation. On April 1, 1894, Pearl Street was closed and the building sold. Today only a plaque marks the place where electrical history was made. Hughes, *Thomas Edison*, p. 34.

44. There was a 50 percent cost overrun on the Pearl Street station.

45. Passer, p. 98.

46. Hughes, *Thomas Edison*, p. 31.

47. By 1890 the Edison Lamp Company was producing nearly a million lightbulbs a year. John A. Garraty, *The New Commonwealth: 1877–1890* (New York: Harper and Row, 1968), p. 81.

48. For Edison's expansion into foreign markets, see Thomas P. Hughes, *Networks of Power: Electrification in Western Society, 1880–1930* (Baltimore: Johns Hopkins University Press, 1983), pp. 47–78.

49. Wyn Wachorst, *Thomas Alva Edison: An American Myth* (Cambridge, Mass.: MIT Press, 1981), p. 23.

50. Ibid., p. 37.

51. Garraty, p. 95.

52. Francis E. Leupp, *A Biography of George Westinghouse* (Boston: Little, Brown, 1918), p. 56.

53. Ibid., p. 48.

54. Passer, p. 129.

55. Ibid., p. 130.

56. Ibid., pp. 130–31.

57. Robert Conot, *A Streak of Luck: The Life and Legend of Thomas Alva Edison* (New York: Seaview Books, 1979), p. 135.

58. Leupp, pp. 132, 131.

59. Lucien Gaulard was a French inventor and John D. Gibbs was his British business partner. Despite a long legal struggle, they were never able to establish a patent claim to the transformer. Bitter and disappointed, Gaulard entered Sainte-Anne Hospital in Paris, an insane asylum, where he died on November 16, 1888. Hughes, *Networks of Power*, p. 94.

60. A. A. Halacsy and G. H. von Fuchs, "Transformer Invented Seventy-five Years

Ago," *American Institute of Electrical Engineers: Power Apparatus and Systems* 80 (June 1961): 121–28.

61. Leupp, p. 138.

62. Hughes, *Networks of Power*, p. 95.

63. The Westinghouse Electric Company was formed on January 8, 1886, with a capital stock of $1 million.

64. Hughes, *Networks of Power*, p. 103.

65. Only six transformers were used in Great Barrington. Ten were sent to Pittsburgh for use in a demonstration plant between the Union Switch and Signal Company and East Liberty, Pennsylvania. William Stanley, "Alternating Current Development in America," *Franklin Institute Journal* 163 (1912): 561–80.

66. Passer, p. 379.

67. Stanley, p. 571.

68. Ibid., p. 572.

69. The Westinghouse Electric Company sold equipment and the patent licenses to local lighting companies that put them into service. License agreement, December 5, 1888, Westinghouse Electric files; Passer, p. 138.

70. Ibid.

71. Conot, p. 254.

72. Passer, p. 138.

73. Conot, p. 254.

74. Westinghouse now had an accurate meter that the customer could read and check. The Edison meter used a chemical meter in which zinc plates were immersed in zinc sulfate. The plates had to be removed and weighed once a month to determine the amount of current used. "Metering Electrical Energy," *Electrical Engineering* 60 (September 1941): 421–22; H. W. Richardson, "The Electric Meter," *General Electric Review* (August 1911): 367.

75. Passer, p. 141.

76. Hughes, *Networks of Power*, pp. 119–20.

77. Leupp, p. 157.

78. Josephson, p. 13.

79. McAuliffe, p. 88.

80. Cheney, pp. 28, 45.

81. Henry G. Prout, *A Life of George Westinghouse* (New York: American Society of Mechanical Engineers, 1921), pp. 300–01, 311.

82. Cheney, pp. 30, 33.

83. Ibid., pp. 47–48.

84. Ibid., p. 49.

85. Josephson, p. 346.

86. Passer, pp. 99–100.

87. "Edison vs. Westinghouse," *Electrical World* (1 January 1887): 7.

88. Ibid., p. 8.

89. Josephson, p. 349.

90. Thomas P. Hughes, "Harold P. Brown and the Executioner's Current: An Incident in the AC-DC Controversy," *Business History Review* 32 (1958): 44.

91. Cheney, p. 44.

92. Lewis B. Stillwell, "Alternating Current Versus Direct Current," *Electrical Engineering* 53, no. 5 (May 1934): 708–9.

93. P. B. Shaw to Charles Batchelor, 27 September 1889. Edison Archives.

94. W. S. Andrews to Edison, 21 May 1887. Edison Archives.

95. W. J. Jenks to Edison, 12 November 1887. Edison Archives.

96. J. H. Herrick to Edison, 30 October 1889. Edison Archives.

97. Westinghouse to Edison, 7 June 1888. Edison Archives.

98. Westinghouse to Edison, 3 July 1888. Edison Archives.

99. Edison to J. H. Herrick, 30 October 1889. Edison Archives.

100. Josephson, p. 346.

101. Cheney, p. 41.

102. "A Warning from the Edison Light Company," pamphlet, New York, February 1888.

103. Josephson, pp. 347–48.

104. "A Warning from the Edison Company," p. 45.

105. Ibid., p. 41.

106. "Edison Predicted It," *New York Evening Sun*, 16 October 1889.

107. Edison to Fredric Harris, 13 December 1893. Edison Archives.

108. Thomas A. Edison, "Scientific," *Public Opinion* (November 1889): 113.

109. Passer, p. 172.

110. A copy of this report can be found in the Sprague Papers, Engineering Societies Library, New York City.

111. Passer, p. 172.

112. Comment of W. S. Andrews, stated in an interview with J. W. Hammond, General Electric Files, quoted in Passer, p. 173.

113. Francis Jehl, 22 April 1913, Hammer Collection, quoted in Conot, p. 468.

114. Reese V. Jenkins et al., eds., *The Papers of Thomas A. Edison* (Baltimore: Johns Hopkins University Press, 1989), vol. 1, p. xxix.

115. Neil Baldwin, *Edison: Inventing the Century* (New York: Hyperion, 1995), p. 210.

116. Josephson, p. 357.

117. *Edison Electric Light v. U.S. Electric Co.*, Circuit Court of the Southern District of N.Y., IV, 257–73: Hearings, June 19, 1890, quoted in Josephson, p. 357.

118. Josephson, p. 357.

CHAPTER THREE The New Electrical Execution Law

1. In the winter of 1607–08, Captain George Kendall became the first man executed in the American colonies. One of the original counselors of the Jamestown colony, he was convicted of mutiny and shot to death. See M. Watt Espy and John Ortiz Smykla, *Executions in the United States, 1608–1987: The Espy File* (Ann Arbor: University of Michigan Press, 1987), p. i.

2. For an account of outings to witness hangings, see W.M.T., "Going to See a Man Hanged," *Fraser's Magazine* 22, no. 128 (August 1840): 150–58.

3. See Charles Duff, *A Handbook on Hanging* (London: Journeyman Press, 1981).

4. Philip Mackey, *Hanging in the Balance: The Anti–Capital Punishment Movement in New York State, 1776–1861* (New York: Garland, 1976), p. 109.

5. See report of the select committee on a resolution directing an inquiry into the propriety of abolishing public executions. State of New York, no. 79. In Senate, April 8, 1835.

6. Thomas J. Davis, *A Rumor of Revolt: The "Great Negro Plot" in Colonial New York* (New York: Free Press, 1985), pp. 80–81.

7. See David Rothman, *The Discovery of the Asylum* (Boston: Little, Brown, 1971).

8. When serving as sheriff of Erie County (Buffalo), New York, future president

Grover Cleveland acted as executioner on two occasions. See Negley K. Teeters, "Scaffolds and Executioners," *Scaffold and Chair*, part 2 (Philadelphia: Temple University Press, 1963), pp. 21–22.

9. With the use of a counterbalance, there was for a given length of a fall a considerable loss of force, according to the formula $(b \times 2 + w)$ 1/w, that is, if the body weighed 140 pounds, and the opposing weight was 300 pounds, the working force for a constant space is only as 1:1.93; if $w = 500$, the force is 1:1:56, a loss considerably reduced. See Alonzo Calkins, "Felonious Homicide: Its Penalty, and the Execution Thereof Judicially," *Papers Read Before the Medico-Legal Society of New York*, second series, revised ed. (New York: W. F. Houghton, 1882), p. 265. Elbridge T. Gerry, "Capital Punishment & Electricity," *North American Review* 149, no. 394 (1889), p. 323.

10. Today researchers refer to this as the brutalization effect. See William J. Bowers and Glenn Pierce, "Deterrence or Brutalization: What Is the Effect of Executions?" *Crime and Delinquency* 26 (1980): 453–84.

11. See Calkins.

12. Report of the select committee on the expediency of a total abolition of capital punishment in the State of New York, *Documents of the Assembly of the State of New York, Fifty-fifth Session*, 1832 (vol. 3, no. 187), p. 16.

13. Ibid., p. 17.

14. Ibid., p. 18.

15. Ibid., p. 20.

16. Ibid., pp. 21, 22.

17. *The Journal of the Assembly of the State of New York, at their Fifty-seventh Session*, 1834, p. 93.

18. *The Revised Statutes of 1828*, III, 659, part IV, chap. 1, title 1, no. 26.

19. *Albany Evening Journal*, 10 March 1834.

20. *Documents of the Assembly of the State of New York, Fifty-seventh Session*, 1834 (vol. 1, no. 79), pp. 93, 350, 355, 365, 387, 398, 410.

21. Report of the select committee on a resolution directing an inquiry into the propriety of abolishing public executions, read in the New York State Senate, 8 April 1835, p. 2. Capital Punishment, New York State, 1832–1872, SLP, p. vi, New York Library Annex.

22. Ibid., p. 3.

23. Ibid., pp. 1–11.

24. *The Journal of the Senate of the State of New York, at their Fifty-eighth Session*, 1835, p. 350. New York became the fifth state to do this, in 1835 (see Act of May 9, 1835, ch. 258, 1835, NY Laws 299), after Connecticut in 1830 (see Act of June 5, 1830, ch. 1, section 147, 1830 Conn. Pub. Acts 284), Rhode Island (see General Assembly "Acts and Resolves" June 1833: 50–51), Pennsylvania in 1834 (see Act of Apr. 10, 1834, Pub. Act No. 127, 1833–34 Pa. Laws 234), and New Jersey in 1835 (see Act of Mar. 5, 1835, 1834–35 N.J. Laws 170).

25. *Documents of the Assembly of the State of New York*, 1839 (vol. 6, no. 378); 1845 (vol. 7, no. 249); 1846 (vol. 6, no. 213); 1847 (vol. 2, no. 95); 1857 (vol. 3, no. 170); 1860 (vol. 3, no. 82).

26. Edward Robb Ellis, *The Epic of New York City: A Narrative History from 1524 to the Present* (New York: Coward-McCann, 1966), p. 251.

27. The final improvement of painkillers came in 1898 when the Bayer Company added an extra chemical to morphine making it able to reach the brain faster, thus producing heroin.

28. Many scientists went to great lengths in experimenting with different methods of execution. On July 24, 1886, the *Medical Record* published an article by Dr. Wooster

Beach. Most current attempts to make executions painless usually contained suggestions for methods other than hanging. Dr. Beach agreed that the popular long-drop method of hanging was ineffective, citing his own study that indicated that in only 5 percent of the cases was the neck actually broken and death instantaneous. Ninety-five percent of the victims suffered "exquisite torture till asphyxia produces insensibility." However, Dr. Beach, unlike most of his contemporaries, did not believe hanging should be entirely abandoned. He cited the work of Dr. Graeme M. Hammond, which suggested that the best way to hang a man was to put a rope around his neck, just above the larynx, and slowly lift him into the air. Hammond claimed (based on experiments of his own in which an assistant tightly constricted his neck with a towel) that hanging wasn't especially painful if done in this way.

    29. "Killing Cattle by Electricity," *Scientific American* 48, no. 12 (1883): 184.

    30. On February 14, 1885, an article in *Scientific American* proposed that prisoners' labor be used to generate electricity to offset the cost of keeping large numbers of prisoners in unproductive idleness. After all, *Scientific American* pointed out, it was nearly impossible to find useful employment for "tramps, drunkards, station house rounders, and the like." Originally, the concept of "hard labor" meant only labor as a punishment. Prisoners were placed in a cell and made to turn a heavy crank, with no attempt to make the activity productive; their efforts produced nothing, and were enforced only for the sake of punishment. The same was true for the treadmill because it was believed that only unproductive labor was a punishment. By forcing prisoners to generate electricity from otherwise unproductive activities for electric lighting and running machinery, "brute energy" could be sold for industrial use. "Electricity for Executing Criminals," *Scientific American* 52, no. 7 (1885): 101.

    31. Ibid.

    32. For more biographical information, see George B. Snow, "Dr. Alfred Porter South-wick," *The Dental Cosmos*, 40 (1898): 597–98.

    33. Th. Metzger, *Blood and Volts: Edison, Tesla, and the Electric Chair* (New York: Autonomedia, 1996), p. 29.

    34. *Buffalo Evening News*, 9 August 1881.

    35. *Electrical World* (24 December 1887): 325.

    36. Herbert J. Bass, *"I Am a Democrat": The Political Career of David Bennett Hill* (Syracuse: Syracuse University Press, 1961), pp. 11–14.

    37. Governor Hill's initial address, January 6, 1885.

    38. "A Long Legal Struggle," *New York Times*, 7 August 1890, p. 2. The legislature approved the formation of a commission by the Laws of 1886, chap. 352, extended by Laws of 1887, chap. 7.

    39. Elbridge Gerry (1744–1814) was a two-term governor of Massachusetts who redrew the state's voting districts to guarantee legislative control by the Republican Party. The term *gerrymandering* originated with him. Elbridge Gerry also served as vice president of the United States in 1813–14.

    40. The organization for children was founded one year after the animal society was founded. It began with a small child named Mary Ellen, who was being abused by her legal guardians. Her situation was discovered by the city, and she was brought to the ASPCA. The children's society was founded soon after. For further information, see Gerry's obituary "Elbridge T. Gerry Dies in 90th Year," *New York Times*, 19 February 1929.

    41. The Gerry Commission asked the following questions in its survey: *"First.* Do you consider the present mode of inflicting capital punishment, by hanging, objectionable?

Please give the reasons for your opinion. *Second.* Were you ever present at an execution, and, if so, will you kindly state details of the occurrence bearing on the subject? *Third.* In your opinion, is there any method known to science which would carry into effect the death penalty in capital cases, in a more humane and practical manner than the present one of hanging? If so, what would you suggest? *Fourth.* The following substitutes for hanging have been suggested to the Commission. What are your views as to each? 1. Electricity. 2. Prussic acid, or other poison. 3. The guillotine. 4. The garrote. *Fifth.* If a less painful method of execution than the present should be adopted, would any legal disposition of the body of the executed criminal be expedient, in your judgment, in order that the deterrent effect of capital punishment might not be lessened by the change? What do you suggest on this head?"

42. "Report of the Commission to Investigate and Report the Most Humane and Practical Method of Carrying Into Effect the Sentence of Death in Capital Cases," transmitted to the legislature of the state of New York, January 17, 1888 (Albany: The Argus Company, 1888), p. 86.

43. R. Ogden Doremus, Clark Bell, J. Mount Bleyer, Chas. F. Stillman, and Frank H. Ingram, "Report of Committee on Best Method of Executing Criminals," *The Medical-Legal Journal* 5, no. 1 (1888): 442–44.

44. Ibid., pp. 427–41.

45. Ibid., pp. 429–32.

46. Ibid., pp. 423–38.

47. Ibid., pp. 438–39.

48. Ibid., p. 441.

49. A. P. Southwick to Thomas Edison, 8 November 1887. Edison Archives.

50. The letter from Edison to A. P. Southwick is missing from the Edison Archives. From the contents of Southwick's second letter and Edison's reply, however, it is clear that Edison opposed capital punishment.

51. A. P. Southwick to Edison, 5 December 1887. Edison Archives.

52. Edison to A. P. Southwick, 9 December 1887. Edison Archives.

53. "The People of The State of New York, ex rel. William Kemmler." Transcripts of the Proceedings before Tracy S. Becker, Esq., Referee, at the office of Hon. W. Bourke Cockran, in the Equitable Building, 120 Broadway, New York City, Monday, July 8, 1889, at 12 M., pursuant to an order of 2 references heretofore made and entered in the above entitled proceeding, p. 372.

54. Three days before the commission's report was released, *Harper's Weekly* published a tongue-in-cheek article entitled "Execution by Electricity." Its author, novelist W. D. Howells, wrote the letter on Christmas Day. Howells, the former editor of *Atlantic Monthly*, served more than two decades as a columnist and critic for *Harper's Weekly*. Howells was an ardent supporter of the doctrine of realism, advocating a "truthful treatment" of "the motives, the impulses, the principles that shape the life of actual men and women." His somewhat cynical approach to the subject of death by electricity was characteristically simple, straightforward, and pragmatic.

Howells noted that the New York legislature would probably pass a bill replacing hanging with execution by electricity that winter. He wrote that there was still much affection for the "good old gallows-tree," a central part of Anglo-Saxon tradition. Howells said many believed that hanging had become an "instrument of torture" and it was our duty to kill by the "humanest method known to science." He sarcastically said many saw little reason that electricity, which had improved their lives in a variety of ways, "should not also be employed to take it away." The advantage, he remarked, of electricity was that the "death spark" could be given to a "victim" while he was asleep. Howells made the deliberately outrageous proposal that the execution take place from the governor's office, where the governor himself could press a button and send the criminal to his death. In celebrated cases, he suggested, the governor could invite distinguished persons to his office or allow some society lady or even a young child to press the button. Howells made one additional suggestion; feigning sympathy for sheriffs, he suggested that the executioner, just as the jury, be drawn from the public. Thus, from time to time, professional men would have the opportunity to prove their support for "the great principle that if it is wrong to take life, a second wrong of the kind dresses the balance and makes it right." Reprinted in *Electrical World* (21 January 1888): 27–28.

55. The commission listed the following crimes as offenses punishable by capital punishment in the ancient Mosaic code: murder; kidnapping; eating leavened bread during the Passover; suffering an unruly ox to be at liberty, if he kill (the ox also to be stoned); witchcraft; bestiality (the beast put to death); idolatry; oppression of widow and fatherless; compounding holy ointment or putting it on any stranger; violation of the Sabbath; smiting of one's father or mother; sodomy; eating the flesh of the sacrifice of peace offerings with uncleanness; eating the fat of offered beasts; eating any manner of blood; offering a child to Moloch; eating a sacrifice of peace-offering; screening an idolater; going after familiar spirits and wizards; adultery (both parties, if female married, and not a bondmaid); incest (three kinds); cursing of parents; unchastity in priest's daughter; blasphemy; stranger coming nigh tabernacle; coming nigh the priest's office; usurping the sacerdotal functions; forbearing to keep Passover, if not journeying; presumption, or despising the word of the Lord; uncleanness, or defiling the sanctuary of the Lord; false pretension to the character of a Divine Messenger; opposition to the decree of the highest judicial authority; unchastity before marriage, when charged by a husband. Commission report, p. 4. The commission recorded the Athenian code of laws established by Draco that were punished by the death penalty. These crimes were: sacrilege; impiety (any open disrespect for religious rites or popular faith); treason; murder, or the attempt to murder; and incendiarism. Commission report, p. 5.

56. Ibid., pp. 5–8.
57. Ibid., p. 14.
58. Ibid., pp. 14–49.
59. Ibid., p. 52.
60. Ibid., pp. 31, 52.
61. Ibid., p. 53.
62. Ibid., p. 54.
63. Ibid., p. 55.
64. Ibid., p. 59.
65. Ibid., pp. 59–63.
66. Ibid., p. 63.
67. Ibid., pp. 69–70.
68. Ibid., p. 78.
69. Ibid.
70. *In re* Kemmler, 7 N.Y.S. 145 (Cayuga County Ct., 1889). OR The People of the State of New York, ex rel. William Kemmler, against Charles F. Durston, as warden of the state prison at Auburn, NY, pp. 366–69. The "too painless" objection to lethal injection is still raised today.
71. "Report of the Commission," p. 79.
72. Ibid., p. 80.

73. Ibid.

74. Ibid., p. 81. The commission did not actually witness these incidents. They were supervised entirely by Dr. Fell.

75. Ibid., pp. 81–84.

76. Ibid., pp. 84–85.

77. Ibid., p. 87. The commission also referenced *The Lancet*, the chief medical journal of London, which expressed its preference for hanging. It believed that hanging was quicker and less painful than destroying life by electricity. And it wrote that all the protest against hanging was against the "bungling way" it had been administered, not against hanging itself.

78. Commission report, pp. 83–84.

79. Ibid., pp. 90, 86.

80. Ibid., pp. 86, 91.

81. Ibid., pp. 88, 89.

82. Ibid., p. 89.

83. According to Section 507 of the Code of Criminal Procedure, violation of this provision was a misdemeanor only.

84. Commission report, p. 90.

85. The Gerry Commission submitted its report on January 17, 1888.

86. "Report of the Legislative Council Committee on Capital Punishment," 3 October 1892, p. 263.

87. "Report of the Legislative Council," p. 263.

88. Ibid.

89. Ibid.

90. Ibid.

91. Ibid.

92. Section 507 of the Code of Criminal Procedure, New York State, signed into law on June 4, 1888.

93. "Legislative Session," *New York Times*, 12 May 1888, p. 3.

94. Craig Brandon, *The Electric Chair: An Unnatural American History* (Jefferson, N.C.: McFarland and Co., 1999), pp. 68, 52.

95. "The Press and Executions," *New York Times*, 6 August 1890.

96. *New York Evening Post*, 8 June 1888, quoted in Michael Madow, "Forbidden Spectacle: Executions, the Public and the Press in Nineteenth-Century New York," *Buffalo Law Review* 43, no. 2 (fall 1995): 544.

97. Madow, pp. 544–45.

98. "The Misdemeanor of the Newspapers," *New York Sun*, 8 August 1890.

99. Godkin was a former editor of *The Nation*, who moved to the *Evening Post* in 1883. Godkin was a critic of the commercialism, sensationalism, and gossip-mongering of the popular press. See Hazel Dicken-Garcia, *Journalistic Standards in Nineteenth-Century America* (Madison, Wis.: University of Wisconsin Press, 1989), pp. 161–203. Godkin first used the term *sensational* in 1869 to describe journalism that played to popular interest in crime, sex, and scandal.

100. Madow, p. 554.

101. "An Unsuccessful Experiment," *Buffalo Express*, 7 August 1890.

102. "Liberty of the Press," *Buffalo Courier*, 7 August 1890.

103. Madow, p. 546.

104. For an analysis of the execution ceremony, see Robert Johnson, *Death Work: A Study of the Modern Execution Process* (Pacific Grove, Calif.: Brooks/Cole, 1990); John Lofland, "The Dramaturgy of State Executions," *State Executions Viewed Historically and*

Sociologically, Patterson Smith Series in Criminology, Law Enforcement and Social Prob
lems, publication no. 170 (London: Chapman and Hall, 1977); Madow, "Forbidden Spectacle."
    105. "Sam Steenburgh's Murders," *New York Times,* 19 April 1878, p. 1.
    106. "Seven Murders Explained," *New York Times,* 20 April 1878, p. 1.
    107. Ibid.
    108. "Sam Steenburgh's Murders."
    109. Commission report, 89.
    110. "Sam Steenburgh's Murders."

CHAPTER FOUR Harold Brown and the "Executioner's Current"

    1. His letterhead read, "Harold Brown. Electrical Engineer, 210 West 54th St. New
York. Designer of Apparatus for Special Purposes. Contract for Arc and Incandescent
Electric Lights and Steam Power. Life Protecting Apparatus for Arc Lights."
    2. Harold P. Brown, "Death in the Wires," *New York Evening Post,* 5 June 1888, p. 7.
    3. Ibid.
    4. Ibid.
    5. Ibid.
    6. "Electrical Death for Dogs," *New York Morning Sun,* 4 August 1888, p. 1.
    7. "Death in the Wires."
    8. "High Potential System Before the Board of Electrical Control of New York," *Elec-
trical Engineer* (August 1888): 360–61.
    9. Some sources imply that Edison approached Brown, but most evidence suggests
that Brown approached Edison.
    10. For a biographical sketch of Arthur Edwin Kennelly, see Vannevar Bush, *A Bio-
graphical Memoir of Arthur Edwin Kennelly 1861–1939,* National Academy of Sciences of
the United States of America, Biographical Memoirs, vol. 22—Fifth Memoir. Presented to
the academy at the autumn meeting, 1940.
    11. Edison Archives. Reel 109 contains more than twenty letters between Brown and
Kennelly describing how Kennelly directed Brown's experiments.
    12. Edison Archives, reel 138/441 and reel 122/882.
    13. T. Carpenter Smith, "Letter to the Editor," *Electrical Engineer* (August 1888): 361.
    14. Ibid.
    15. Ibid., p. 362.
    16. Ibid.
    17. H. M. Byllesby, "Letter to the Editor," *Electrical Engineer* (August 1888): 367.
    18. Ibid., p. 368.
    19. "High Potential System."
    20. Ibid.
    21. "Physiological Tests with the Electric Current," *Electrical World* (11 August 1888):
369.
    22. "Died for Science's Sake," *New York Times,* 31 August 1888, p. 6.
    23. Ibid.
    24. Ibid.
    25. "Physiological Tests with the Electric Current," pp. 369–70.
    26. "Died for Science's Sake," p. 8.
    27. Harold P. Brown, "The Comparative Danger to Life of Alternating and Continu-
ous Currents" (self-published pamphlet, 1889), p. 12.
    28. State of New York, "The People of the State of New York, Ex. Rel. William Kemm-

ler, Appellant, Against Charles F. Durston, Agent and Warden of Auburn Prison, Respondent," 2 vols. Bound in Court of Appeals, 1847–1911 (Buffalo, N.Y.: 1890), p. 30 (#117).

29. "Electrical Death for Dogs."

30. Ibid.

31. P. H. Van der Weyde, "The Comparative Danger of Alternate vs. Direct Currents," *Electrical Engineer* (September 1888): pp. 451–54.

32. Ibid., p. 451.

33. Ibid.

34. Ibid., p. 452.

35. Ibid., p. 454.

36. George Westinghouse, Jr., to the *New York Evening Post*, 10 December 1888.

37. "Electricity as a Life-Taker," *New York Morning Sun*, 4 November 1888.

38. Ibid.

39. The organization was founded in June 1866. For a history of New York medical societies, see James J. Walsh, *History of Medicine in New York: Three Centuries of Medical Progress* vol. 3 (New York: National Americana Society, 1919). The subject of executions was first introduced by the eminent French scientist Ambrose Tardieu, who had submitted a report entitled "Diagnosis of Hanging" in 1870. Dr. Alonzo Calkins had presented a paper before the society in September 1873; Calkins advocated the abolition of death by hanging after discussing various other methods of execution. The discussion was renewed before the society on April 3, 1878, by Professor J. H. Packard of Philadelphia, who strongly encouraged the elimination of hanging. He recommended a painless death by the inhalation of sulfuric oxide gas, claiming it was the most humane method of execution available.

40. The other members of the committee were Dr. J. Mount Bleyer, Professor R. Ogden Doremus, and Dr. Frank H. Ingram, formerly assistant superintendent of the Blackwell's Island Insane Asylum. "A Death Helmet for Them," *New York World*, 15 November 1888.

41. Ibid.

42. "Death Current Experiments at Edison's Laboratory," *Electrical World* (15 December 1888): 312–13. Reprinted in Harold Brown, "Death Current Experiments at the Edison Lab," *Medico-Legal Journal* 6 (1888): 386–89.

43. "The Electrical Death Penalty," *Electrical World* (12 January 1889): 17.

44. "Electrical Execution: The Substitute for the Gallows Not Perfected," *New York Advertiser*, 3 August 1888.

45. Ibid.

46. Peterson invited Edison to attend the meeting and the annual banquet that followed the proceedings, but Edison declined the offer. Peterson to Edison, 11 December 1888. Edison Archives.

47. Report of the Committee of the Medico-Legal Society on the Best Method of Execution of Criminals by Electricity. Edison Archives, reel 122/981–83.

48. "Bound to the Death Chair: A Highly Scientific Device for Electrical Executions," *New York World*, 13 December 1888. Other information is available in the Edison Archives, reel 146/362.

49. The dissenter, Henry Guy Carleton of the newspaper *The World*, presented his opinion at the meeting. Carleton's paper was a long, detailed account of electrical execution, including many of the technical questions involving voltage, resistance, and placement of the electrodes. The most interesting part of his paper was his assertion that a switch to electricity would not eliminate the barbarity of executions. He cited the French practice of keeping the day designated for the execution unknown to the condemned, who

was awakened from his sleep and executed in the shortest time possible. He went on to say that everyone was condemned to death, yet because the day of our deaths have not been fixed, we live most of our lives without dreading our demise. Consequently, Carleton argued, the first point of barbarity was fixing the date of the execution. He suggested that the most humane method would be to fill his cell with an odorless gas at a time not disclosed to the condemned, and while he sleeps, allowing the prisoner to die a painless and unexpected death. He approvingly quoted Professor C. F. Brackett of Princeton College as saying that no one would ever die in the electric chair; Brackett claimed that New York would never have the "nerve" to carry into effect such an "outrage." He was dead wrong. "Execution by Electricity: Alleged Faults and Defects of the New Death Penalty," *New York Press*, 5 January 1889.

50. George Westinghouse, Jr., to the *New York Evening Post*.

51. Brown to the Thomson-Houston Electric Company, 14 March 1889. Quoted in "For Shame, Brown!" *New York Sun*, 25 August 1889, p. 6.

52. Harold Brown to the *New York Tribune*, 18 December 1888. Edison Archives, reel 122/994.

53. Edison Archives, reel 126/42. Also, Brown to George Westinghouse, Jr., *Electrical Engineer* (4 April 1889).

54. Brown to Kennelly, 4 August 1888. Edison Archives.

55. *Electrical Engineer* 8 (1889): 74.

56. Ibid.

57. Brown, "The Comparative Danger to Life," p. 10.

58. Ibid.

59. They were also published in "Physiological Test with the Electric Current," *Electrical World* 12 (11 August 1888): 69–72.

60. Brown, "The Comparative Danger to Life," p. 11.

61. Ludwig Gutmann, "A Review of Mr. Harold P. Brown's Experiments," letter to the editor, *Electrical World* 14 (13 July 1889): 25–26.

62. Ibid.

63. Ibid., pp. 25–26.

64. "The Constitutionality of Execution by Electricity," *Electrical World* 14, no. 7 (1889): 105–8.

65. Ibid., pp. 105, 106.

66. Ibid.

67. Ibid.

68. Ibid.

69. Ibid., p. 107.

70. Ibid.

71. Ibid.

72. Ibid.

73. Ibid., p. 108.

74. Ibid.

75. Ibid.

76. Ibid.

77. Brown offered $500 as a reward for the return of his letters and a conviction of the thief (*New York Times*, 8 May 1889, p. 22). He also claimed that some of the letters were forgeries, arguing that he often "made carbon copies of my letters, but always before signing them" (Edison Archives, reel 146/452). "His Desk Robbed," *New York Journal*, 4 September 1889.

78. "For Shame, Brown!" *New York Sun*, 25 August 1889, p. 6.
79. Ibid.
80. Ibid.
81. Ibid.
82. For the test results, see Louis Duncan and W. F. C. Hasson, "Some Tests on the Efficiency of Alternating Current Apparatus," *Electrical Engineer* (2 April 1890): 158–60. Also, Brown to Westinghouse.
83. "For Shame, Brown!"
84. Ibid.
85. Edison Archives, reel 126/45.
86. "For Shame, Brown!"
87. Ibid.
88. "Mr. Harold P. Brown," *Electrical World* 14 (31 August 1889): 153–54. Also, Edison Archives, reel 146/485.
89. Elbridge T. Gerry, "Capital Punishment by Electricity," *North American Review* 149, no. 394 (1889): 321, 322.
90. Ibid., 322–23.
91. Ibid., pp. 323–24. A similar report written by Edison is available in the Edison Archives, reel 104.
92. Harold P. Brown, "The New Instrument of Execution," *North American Review* 149, no. 396 (1889): 586–93.
93. Ibid., pp. 586–87.
94. Ibid., pp. 588–93.
95. Ibid., pp. 592–93.
96. Ibid., p. 593.
97. Thomas Edison to Lloyd Bryce, Esq. (October 1889). Edison Archives, reel 125/639.
98. "The Brush Electric Co. Sells Out to the Thomson-Houston Electric Co.," *Electrical Engineer* (November 1889): 498.
99. Thomas A. Edison, "The Dangers of Electric Lighting," *North American Review* 149, no. 396 (1889): 625–34.
100. Ibid., p. 626.
101. Ibid., pp. 626–30.
102. Ibid., pp. 630, 632–34.
103. Although the article did not carry a byline, judging from its content and position in the journal, it is clear that the Westinghouse Company was its author. "Harold P. Brown Again," *Electrical Engineer* (November 1889): 498.
104. See Deborah W. Denno, "Is Electrocution an Unconstitutional Method of Execution? The Engineering of Death over the Century," *William and Mary Law Review* 35, no. 2 (winter 1994): 591–692.

CHAPTER FIVE   "I'll Take the Rope": The Life, Crime, and Trial of William Kemmler

1. *Auburn Citizen*, 5 August 1890, pp. A1, A8.
2. Matilda's maiden name was Lesler. "Murderer Kemmler," *Buffalo Daily Times*, 30 March 1889, p. 1, cols. 7, 8.
3. "Kemmler in Jail," *Buffalo Evening News*, 1 April 1889, p. 1.
4. "It Is Murder: Mrs. Hort Dies This Morning of the Wounds Her Husband Gave Her," *Buffalo Evening News*, 30 March 1889, p. 1.

5. "Kemmler's Crime: The Brutal Murder of Mrs. Tillie Ziegler in Buffalo Last Year," *World*, 8 August 1890, p. 2.

6. Ibid.

7. Ibid.

8. "Busy Counting Eggs: Within Thirty Feet of Where a Woman Was Hacked to Pieces," *Buffalo Courier*, 2 April 1889, p. 1.

9. "He Pleads Guilty," *Buffalo Daily Times*, 1 April 1889, p. 1.

10. Matilda Ziegler was buried by the Crowley Brothers in Forest Lawn on April 3, 1889. Her father was extremely emotional during the funeral and broke down completely at times. "Matilda Buried," *Buffalo Evening News*, 3 April 1889.

11. "It Is Murder."

12. "The Jury Is Ready: Twelve Men Who Will Sit in the Box," *Buffalo Express*, 7 May 1889, p. 1.

13. Ibid.

14. "For His Life: Murderer Kemmler on Trial in the Supreme Court for Killing His Wife," *Buffalo Evening News*, 7 May 1889, p. 1.

15. For a brief discussion of the history of death-qualified juries, see Hugo Adam Bedau, *Death Is Different: Studies in the Morality, Law, and Politics of Capital Punishment* (Boston: Northeastern University Press, 1987), pp. 147–48. See also Craig Haney, ed., "Special Issue: Death Qualifications," *Law and Human Behavior* 8 (1984): 1–195.

16. The jurymen were Frank Genin, a Lancaster plow-chaser; Frederick Lansell, a steady employee of the Nelson Rendering Works; George Meyer, a Boston farmer; Daniel Chadderdon, an elderly farmer from Marilla; C. J. Swyers, an agriculturist from Alden; William H. Henshaw of Marilla; Jacob Nehrbass, a Newstead farmer; Hiram Darling, an Alden agriculturist; Jerome Brown, a barber from Eden; Augustus Melins of West Seneca; Hiram Emerson, a carpenter from Concord; and John Thuman of Buffalo's Seventh Ward.

17. "The Jury Is Ready."

18. "He Wants to Hang," *Buffalo Express*, 30 March 1889, p. 1.

19. "The Murder Trial," *Buffalo Evening News*, 7 May 1889, p. 1.

20. "For His Life."

21. Ibid.

22. Ibid.

23. *The People of the State of New York v. William Kemmler* (Buffalo, N.Y.: Printing House of James D. Warren's Sons, 1889), Certified, Charles A. Orr, 1 July 1889, pp. 9–15 (#35–59).

24. Ibid., pp. 16–22 (#60–85).

25. Ibid., pp. 22–30 (#86–119).

26. Ibid., pp. 30–36 (#119–141).

27. Ibid., pp. 36–40 (#142–159).

28. "For His Life."

29. *People v. William Kemmler*, p. 44 (#172–175).

30. Ibid., p. 45 (#176–179).

31. Ibid., p. 47 (#186).

32. Ibid., pp. 48–52 (#188–207).

33. Ibid., pp. 53–59 (#208–236).

34. Ibid., pp. 64–68 (#255–270).

35. "For His Life."

36. *People v. William Kemmler*, pp. 79–84 (#314–332).

37. Ibid., pp. 88–90 (#348–357).

38. Ibid., p. 92 (#366).

39. "Jealous Kemmler: The Defense of the Man Who Chopped His Paramour to Death," *Buffalo Evening News*, 8 May 1889, p. 1.

40. Ibid.

41. *People v. William Kemmler*, pp. 104–105 (#413–420).

42. Ibid., pp. 108–14 (#428–455).

43. Ibid., pp. 115–16 (#458–461).

44. Ibid., pp. 116–17 (#461–464).

45. Ibid., pp. 118–20 (#470–478).

46. Ibid., pp. 120–21 (#479–480).

47. For a discussion about the difference between motive and criminal intent, see Justin Miller, "The Mental Elements in Crime," *Criminal Law*, ch. 5 (St. Paul, Minn.: West Publishing Co., 1934), pp. 52–76.

48. *People v. William Kemmler*, pp. 122–23 (#485–491).

49. Ibid., pp. 124–25 (#493, 495, 496).

50. Ibid., p. 127 (#500, 504).

51. "Jealous Kemmler."

52. *People v. William Kemmler*, pp. 132–37 (#524–547).

53. Ibid., p. 138 (#548).

54. Ibid., pp. 143–49 (#568–592).

55. Ibid., p. 150 (#597–598).

56. Ibid., p. 162 (#647).

57. "Kemmler in Jail."

58. *People v. William Kemmler*, p. 153 (#609).

59. Ibid., pp. 156, 157 (#621, 625).

60. Ibid., pp. 159, 160 (#633–634, 636).

61. Ibid., pp. 161–63 (#640–650).

62. "Jealous Kemmler."

63. Ibid.

64. For a discussion of alcoholic insanity as defined during this period, see H. W. Mitchell, "Types of Alcoholic Insanity, with Analysis of Cases," *American Journal of Insanity* (October 1904): 251–74.

65. "Jealous Kemmler."

66. *People v. William Kemmler*, pp. 165–66 (#656–657, 660–662).

67. Ibid., pp. 167, 168–69 (#664, 671–672, 675).

68. Ibid., pp. 171, 172–73 (#680, 683–690).

69. Ibid., pp. 176, 177 (#700, 702, 706).

70. Ibid., pp. 178, 179, 181 (#711, 715, 720).

71. Ibid., pp. 185, 189 (#736, 755).

72. "Jealous Kemmler."

73. *People v. William Kemmler*, p. 192 (#767).

74. Ibid., pp. 197, 199, 200 (#785–786, 794, 797).

75. Ibid., pp. 201–3 (#800–809).

76. Ibid., pp. 203–8 (#810–828).

77. "Jealous Kemmler."

78. *People v. William Kemmler*, pp. 210, 211 (#836, 840).

79. "Jealous Kemmler."

80. *People v. William Kemmler*, pp. 225–28 (#897–909).

81. Ibid., p. 228 (#909–911).

82. Ibid., pp. 231–35 (#921–938).

83. "Tales of Drunkards: They Are Expected to Help Kemmler Out," *Buffalo Express*, 9 May 1889, p. 1.

84. *People v. William Kemmler*, pp. 235–304 (#939–1213).

85. Ibid., pp. 304–5, 306 (#1215–1217, 1220).

86. Ibid., pp. 309–34 (#1233–1334).

87. Ibid., pp. 336–37, 339–40 (#1341–1347, 1355–1359).

88. Ibid., pp. 343, 344–46 (#1369–1371, 1373–1382).

89. Ibid., pp. 347–50 (#1385–1397).

90. Ibid., pp. 350–52 (#1398–1405).

91. Ibid., p. 353 (#1411).

92. Ibid., p. 354 (#1412–1415).

93. Ibid., p. 356 (#1420).

94. Ibid., pp. 356–57 (#1421–1425).

95. Ibid., pp. 367–68 (#1426–1428).

96. Ibid., pp. 358–61 (#1429–1443).

97. Ibid., pp. 361–67 (#1443–1467).

98. Between 460 and 370 B.C., Hippocrates drafted the first detailed account of tuberculosis, which was known as phthisis. He correctly referred to tuberculosis as a disease of the lungs. At Hippocrates' time, tuberculosis was believed to be inherited. This assumption would continue until Kemmler's day. In 1882 Robert Koch discovered the tubercle bacillus, the bacteria that causes tuberculosis. Two years later he published a complete paper on the etiology of tuberculosis in which he provided evidence that tuberculosis, or consumption as it was known at the time, was not hereditary. However, much of the public and the medical community maintained their belief in a hereditary link. By the time of Kemmler's trial in 1889, experiments were still being conducted to find evidence of the passage of the disease from mother to fetus. Richard M. Burke, *An Historical Chronology of Tuberculosis* (Springfield, Ill.: Charles C Thomas, 1938).

99. *People v. William Kemmler*, p. 368 (#1469, 1470).

100. Ibid., p. 369 (#1475).

101. For up-to-date information on syphilis see National Institutes of Health, www.niaid.nih.gov/factsheets/stdsyph.gov.

102. "Kemmler's Fate: The Murderer Was Jealous of His Paramour and 'Yellow' De Bella," *Buffalo Evening News*, 9 May 1889, p. 1.

103. *People v. William Kemmler*, p. 378 (#1508).

104. "The Case Closed: All the Evidence Presented in the Kemmler Murder Trial," *Buffalo Evening News*, 9 May 1889, p. 3.

105. *People v. William Kemmler*, pp. 380, 369 (#1519, 1475).

106. Ibid., pp. 381–82 (#1523–1527).

107. Ibid., pp. 382–83 (#1527–1530).

108. Ibid., pp. 387–88 (#1544–1549).

109. Ibid., pp. 388–91 (#1550–1563).

110. Ibid., pp. 391–93 (#1564–1568).

111. Ibid., pp. 393–97 (#1569–1584).

112. Ibid., pp. 397–98 (#1585–1591).

113. Ibid., pp. 397–400 (#1585–1598).

114. *People v. Ira Stout*, 101 NY Parker's Criminal Report, April 1858, pp. 670–81.

115. *People v. William Kemmler*, pp. 397–400 (#1585–1598).

116. Ibid., pp. 402–6 (#1607–1621).

117. For a social history of the origin of the McNaughtan Rules, see Richard Moran, *Knowing Right from Wrong: The Insanity Defense of Daniel McNaughtan* (New York: Free Press, 1981).

118. *People v. William Kemmler*, pp. 406–7 (#1622–1624).

119. Ibid., pp. 407–9 (#1625–1635).

120. Ibid., pp. 410–11 (#1636–1643).

121. Ibid., pp. 411–12 (#1641–1644).

122. Ibid., pp. 412–13 (#1646–1649).

123. Amasa J. Parker, Jr., *The New York Code of Civil Procedure*, revised by Albert J. Danaher, 3rd ed. (Albany, N.Y.: Banks & Co., 1903), p. 228.

124. *People v. William Kemmler*, p. 413 (#1650–1651).

125. Ibid., pp. 414–15 (#1652–1659).

126. Ibid., pp. 416–17 (#1661–1665).

127. Ibid., pp. 418–19 (#1670–1673).

128. "For His Life: Murderer Kemmler on Trial in Supreme Court for Killing His Wife," *Buffalo Evening News*, 7 May 1889, p. 1.

129. *People v. William Kemmler*, p. 419 (#1674–1675).

130. Ibid., p. 420 (#1676).

131. Ibid., pp. 420–22 (#1679–1685).

132. Ibid., pp. 422–23 (#1686–1691).

133. Ibid., pp. 423–33 (#1692–1731).

134. Ibid., pp. 433–40 (#1732–1758).

135. Ibid., pp. 440–41 (#1759–1766).

136. Ibid., pp. 442–46 (#1767–1781).

137. Ibid., pp. 446–47 (#1781–1784).

138. Ibid., pp. 447–48 (#1785–1788).

139. Ibid., p. 448 (#1789–1798).

140. Ibid., pp. 448–50 (#1789–1799).

141. "To Die at Auburn," *Buffalo Courier*, 12 May 1889, p. 1.

142. "His Doom: Murderer Kemmler Sentenced by Judge Childs to Die by Electricity," *Buffalo Evening News*, 14 May 1889, p. 1.

143. Ibid.

144. Ibid.

145. "Kemmler in Jail," *Buffalo Commercial Advertiser*, 18 May 1889, p. 3.

146. "Attorney Hatch Expects to Appeal," *Buffalo Evening News*, 23 May 1889, p. 1.

147. "Kemmler at Auburn," *Buffalo Commercial Advertiser*, 24 May 1889, p. 1.

148. "Murderer's Song: Murderer Composed and Sung It in Prison—Interesting Memento of the First Electric Victim," *Buffalo Evening News*, 27 May 1889.

149. "Kemmler in Prison," *Buffalo Commercial Advertiser*, 27 May 1889, p. 1.

CHAPTER SIX   Unusually Cruel Punishment

1. This writ is guaranteed by U.S. Constitution, Article I, Section 9, and by state constitutions.

2. James McGurrin, *Bourke Cockran: A Free Lance in American Politics* (New York: Charles Scribner's Sons, 1948), p. 95.

3. Andy Logan, *Against the Evidence: The Becker-Rosenthal Affair* (New York: McCall, 1970), pp. 2–257.

4. Robert McElroy, ed., *In the Name of Liberty: Selected Addresses by William Bourke Cockran* (New York: G. P. Putnam's Sons, 1925), p. v.

5. Logan, p. 257.

6. McGurrin.

7. "Why the Hon. Bourke Cockran Took Up the Kemmler Case," *Electrical Review* 15 (16 October 1889): 10.

8. "History of the Dead Murderer: The Wealthy Corporation that Fought to Save Kemmler from His Doom," *New York Tribune*, 7 August 1890.

9. State of New York, "The People of the State of New York, Ex. Rel. William Kemmler, Appellant, Against Charles F. Durston, Agent and Warden of Auburn Prison, Respondent," 2 vols. Bound in Court of Appeals, 1847–1911 (Buffalo, N.Y.: 1890), p. 1045 (#4167).

10. Ibid.

11. Craig Brandon, *The Electric Chair: An Unnatural American History* (Jefferson, N.C.: McFarland and Co., 1999), p. 112.

12. *People ex. rel. Kemmler v. Durston*, pp. 1045–47 (#4168–4173).

13. Th. Metzger, *Blood and Volts: Edison, Tesla, and the Electric Chair* (New York: Autonomedia, 1996), p. 30.

14. "Is Electrical Execution a Cruel and Unusual Form of Punishment?" *Electrical World* 14, no. 3 (1889): 43–44.

15. A person on whose behalf a legal action is brought.

16. *People ex. rel. Kemmler v. Durston*, pp. 2–6 (#5–23).

17. Ibid., p. 2 (#4–7).

18. Ibid., p. 7 (#24).

19. Ibid., pp. 10–11 (#36–43).

20. Ibid., pp. 13, 15 (#48, 58).

21. Ibid., pp. 13–14 (#50–53).

22. Ibid., pp. 18–19 (#71–73).

23. Ibid., pp. 20–21 (#76–81).

24. Ibid., p. 21 (#83).

25. *New York Evening Post*, 14 May 1889.

26. *People ex. rel. Kemmler v. Durston*, p. 23 (#89).

27. Ibid., pp. 23–24 (#91–93).

28. Ibid., p. 24 (#94).

29. Ibid., pp. 24–25 (#95–98).

30. Ibid., p. 26 (#100).

31. Ibid., pp. 26–27 (#101–106).

32. Ibid., pp. 29–30 (#112–118).

33. Ibid., p. 32 (#126).

34. Ibid., pp. 35, 36 (#137, 141).

35. Ibid., p. 39 (#152).

36. Ibid., p. 39 (#155).

37. Ibid., pp. 44–45 (#172–179).

38. Ibid., p. 47 (#184–187).

39. Ibid., p. 48 (#188–189).

40. Ibid., p. 54 (#213).

41. Ibid., p. 57 (#224).

42. Ibid., p. 57 (#227).

43. Ibid., pp. 58–59 (#228–235).

44. Ibid., p. 67 (#267).

45. Ibid., p. 95 (#379).

46. Ibid., p. 99 (#392–393).

47. "Investigating Electrocution," *New York Herald*, 10 July 1889.

48. "Is Electrical Execution a Cruel and Unusual Form of Punishment?" p. 43.
49. *People ex rel. Kemmler v. Durston*, pp. 107–20 (#425–477).
50. Ibid., p. 123 (#491).
51. Ibid., pp. 124–25 (#493–496).
52. Ibid., pp. 125–26 (#499–501).
53. "Electric Executions: Testing the Wheatstone Bridge at the Edison Laboratory," *New York Times*, 13 July 1889, p. 3.
54. Ibid.
55. Ibid., pp. 1, 3.
56. Daniel Gibbens died the following year. For more on his life see "Obituary: Daniel L. Gibbens," *New York Tribune*, 3 May 1890, p. 2.
57. *People ex rel. Kemmler v. Durston*, pp. 151–240 (#630–958).
58. Ibid., pp. 164–65 (#656).
59. "Power of Electricity," *New York Times*, 16 July 1889, p. 1.
60. *People ex rel. Kemmler v. Durston*, 208–23 (#830–891).
61. Ibid., p. 243 (#968).
62. Ibid., pp. 240–80 (#959–1616).
63. Ibid., p. 251 (#1000).
64. "The Kemmler Case," *New York Advertiser*, 17 July 1889.
65. *People ex rel. Kemmler v. Durston*, pp. 345–68 (#1379–1470).
66. Ibid., p. 372 (#1485, 1486).
67. Ibid., pp. 381–98 (#1521–1591).
68. Ibid., pp. 398–451 (#1588–1802).
69. A. D. Rockwell, *Rambling Recollections: An Autobiography* (New York: Paul B. Hoeber, 1920), p. 222.
70. *People ex rel. Kemmler v. Durston*, pp. 569–81 (#2276–2322).
71. Ibid., p. 608 (#2430).
72. Harold P. Brown to Mr. Samuel Insull. Edison Archives, reel 126/50.
73. "Menlo Park's Wizard," *New York Herald*, 23 July 1889.
74. *People ex rel. Kemmler v. Durston*, pp. 624–28 (#2490–2503).
75. Ibid., pp. 631–36 (#2515–2537).
76. Ibid., pp. 646, 647 (#2575, 2579).
77. Ibid., pp. 648–49 (#2585–2590).
78. Ibid., p. 650 (#2592).
79. Ibid., pp. 666–83 (#2592–2726).
80. Ibid., pp. 694–96 (#2767–2778).
81. Ibid., p. 698 (#2784–2786).
82. Ibid., pp. 718–25 (#2863–2892).
83. Ibid., pp. 878–82 (#3504–3522).
84. Ibid., p. 893 (#3564–3566).
85. Ibid., p. 952 (#3802).
86. Ibid., pp. 939–41 (#3751–3758).
87. Ibid., p. 973 (#3886).
88. Ibid., pp. 973–78, 984–85 (#3883–3906, 3927–3934).
89. Ibid., pp. 986–88, 1003 (#3935–3946, 4003).
90. Ibid., p. 1006 (#4018).
91. Ibid., pp. 1007–9 (#4019–4029).
92. Ibid., pp. 1016–24 (#4056–4090).

CHAPTER SEVEN Neither Cruel nor Unusual Punishment

1. "To Kill by Electricity: Mr. Brown Visits Auburn to Look the Machine Over," *New York Times*, 6 August 1889, p. 6.
2. "For Shame, Brown!" *New York Sun*, 25 August 1889, p. 6.
3. *In re Kemmler* 7 N.Y.S. 145 (Cayuga County Ct., 1889), pp. 147, 148.
4. State of New York, "The People of the State of New York, Ex. Rel. William Kemmler, Appellant, Against Charles F. Durston, Agent and Warden of Auburn Prison, Respondent," 2 vols. Bound in Court of Appeals, 1847–1911 (Buffalo, N.Y.: 1890).
5. *In re Kemmler*, p. 148.
6. Ibid., pp. 148, 149.
7. Ibid., p. 149.
8. Ibid., p. 150.
9. Ibid., pp. 150, 151.
10. Ibid., pp. 151–52.
11. Ibid., p. 152.
12. *People ex. rel. Kemmler v. Durston* 7 N.Y.S. 813 (Sup. Ct., 1889), p. 813.
13. Ibid., pp. 814–15.
14. Ibid., p. 815.
15. Ibid., pp. 816, 817.
16. Ibid., p. 818.
17. Ibid., p. 818.
18. *People ex. rel. Kemmler v. Durston*, Brief on Behalf of the Relator, pp. 3, 4.
19. Ibid., pp. 5, 6, 7.
20. Ibid., p. 9.
21. Ibid., pp. 10, 13.
22. Ibid., pp. 14, 16, 17.
23. Ibid., pp. 21, 23.
24. Ibid., p. 23.
25. Ibid., pp. 24, 25.
26. Ibid., pp. 26, 31.
27. Ibid., pp. 36, 37.
28. Ibid., p. 37.
29. *People ex rel. Kemmler v. Durston*, Brief of Attorney-General for Respondent, pp. 7, 9, 17.
30. Ibid., pp. 19–20, 23.
31. Ibid., p. 28.
32. Ibid., pp. 29, 30.
33. Ibid., p. 32.
34. Ibid., pp. 35, 39.
35. Ibid., pp. 52, 54.
36. Ibid., pp. 55, 56, 63.
37. Ibid., p. 67.
38. *People ex. rel. Kemmler v. Durston* 119 N.Y. 569 (Ct. of Appeals 1889), p. 577.
39. Ibid., pp. 578, 579.
40. *The People of the State of New York v. William Kemmler* 119 N.Y. 580 (Ct. of Appeals 1890), p. 580.
41. Ibid., p. 585.
42. *People v. Stout* 3 Park. C. R., pp. 670, 674.

43. *The People of the State of New York v. William Kemmler* 119 N.Y. 580 (Ct. of Appeals 1890), p. 585.

44. Ibid., p. 586.

45. Ibid.

46. Ibid., p. 587.

47. Quoted in Arnold Beichman, "The First Electrocution," *Commentary* (May 1963): 414.

48. "Kemmler Makes His Will: A Belief That He Will Meet His Fate To-Morrow," *New York Times*, 29 April 1890, p. 8.

49. Ibid.

50. "Hope Again for Kemmler," *New York Times*, 30 April 1890.

51. Ibid.

52. Ibid.

53. Ibid.

54. Ibid.

55. Ibid.

56. Ibid.

57. Ibid.

58. "Who Is Kemmler's Friend?" *New York Tribune*, 1 May 1890, p. 4.

59. Ibid.

60. *New York Daily Tribune*, 24 May 1890, p. 1.

61. "Kemmler's Fight for Life," *New York Times*, 6 May 1890.

62. *New York Daily Tribune*, 24 May 1890, p. 1.

63. Ibid.

64. "Kemmler's Fate," *Electrical Review*, 21 June 1890, p. 3.

65. Ibid.

66. "Is It the Dynamo Again? Rushing Through the Bill to Abolish Capital Punishment," *New York Daily Tribune*, 2 May 1890, p. 1.

67. Ibid.

68. Craig Brandon, *The Electric Chair: An Unnatural American History* (Jefferson, N.C.: McFarland, 1999), p. 152.

69. "Is It the Dynamo Again?

70. "Death of an Uncalled-for Measure: The Bill to Abolish Capital Punishment Killed in Committee," *New York Daily Tribune*, 6 May 1890, p. 2.

71. "Kemmler's Fight For Life," p. 9.

72. Ibid.

73. Ibid.

74. Ibid.

75. Ibid.

76. Ibid.

77. Ibid.

78. Ibid.

79. Ibid.

80. Bryan A. Garner, ed., *Black's Law Dictionary: Deluxe Thumb-Index* (St. Paul: West Group: 1999), p. 1610.

81. *In re Kemmler* 136 U.S. 436 (1890), p. 437.

82. Matthew Hale, "The Kemmler Case," *Albany Law Journal* (10 May 1890): 364–67.

83. "Current Topics," *Albany Law Journal* (10 May 1890): 361.

84. Ibid., pp. 361–62.

85. "The Execution of Kemmler," *New York Tribune*, 9 August 1890, p. 1.

86. Ibid.

87. *In re Kemmler* 136 U.S. 436 (1890).

88. Ibid., p. 437.

89. "Kemmler Loses: The U.S. Supreme Court Decides Against Him," *Electrical Review* (31 May 1890): 1.

90. For an excellent discussion of the court's decision see, Deborah W. Denno, "Is Electrocution an Unconstitutional Method of Execution? The Engineering of Death over the Century," *William and Mary Law Review* 35, no. 2 (winter 1994): 551–692.

91. Ibid., p. 590.

92. Ibid.

93. *Robinson v. California* 370 U.S. 660 (1962).

94. William Blackstone, *Commentaries on the Laws of England*, 4 vols., 1765–69 (Chicago: University of Chicago Press, 1979).

95. *Wilkerson v. Utah* 99 U.S. 130 (1878).

96. *In re Kemmler* 136 U.S. 436 (1890), p. 447.

97. Denno, pp. 591–92.

98. *In re Kemmler* 136 U.S. 436 (1890), p. 448.

99. Ibid., pp. 448–49.

100. Ibid., p. 449.

101. "The Kemmler Case," *New York Times*, 24 May 1890.

102. Ibid.

103. Ibid.

104. "Electrical Execution—Kemmler vs. Durston," *Electrical Engineer* (2 July 1890): 25.

105. Ibid.

106. "Kemmler Resentence," *Buffalo Evening News*, 13 July 1890, p. 1.

107. "Kemmler Sentenced to Die by Electricity During the Week of Aug. 4," *Buffalo Evening News*, 12 July 1890, p. 1.

108. Ibid.

109. Ibid.

110. "Kemmler Back," *Auburn Daily Advertiser*, 14 July 1890, p. 1.

CHAPTER EIGHT Kemmler's Legacy:
The Search for a Humane Method of Execution

1. Michael Madow, "Forbidden Spectacle: Executions, the Public and the Press in Nineteenth-Century New York," *Buffalo Law Review* 43, no. 2 (fall 1995): 555.

2. "The Electrical Execution," *New York Times*, 8 July 1891, p. 4, col. 2.

3. " 'Let Her Go!' Shouted McElvaine and Instantly Died," *World*, 8 February 1892, p. 18.

4. In 1906, the citizens of Sing Sing no longer wished to be known for its prison and changed the name of their town to Ossining. The name Sing Sing is derived from the Indian "ossine ossine," which means "stone upon stone." For a history of the early days of Sing Sing see Orlando F. Lewis, *The Development of American Prisons and Prison Customs, 1776–1845* (Albany: New York Prison, 1922), pp. 107–11.

5. "Two Shocks Were Needed," *New York Times*, 9 February 1892.

6. Ibid.

7. Ibid.

8. Ibid.

9. In the course of executing five men in Sing Sing—four in a single day—the decision was made to change one of the electrodes from the base of the spine to the calf of the leg.

10. "Two Shocks Were Needed."

11. Ibid.

12. Ibid.

13. Ibid.

14. Ibid.

15. Ibid.

16. Edison Archives 146/970.

17. "Two Shocks Were Needed."

18. "Edison on Executions," *Newark Call,* 14 February 1890.

19. Ibid.

20. "Two Shocks Were Needed."

21. "Experimental Executions," *Post Express,* 9 February 1892.

22. See *Malloy v. South Carolina,* 237 U.S. 180, 185 n. 1 (1915).

23. Ibid.

24. Deborah W. Denno, "Getting to Death: Are Executions Constitutional?," *Iowa Law Review* 82, no. 2 (January 1997): 407–8.

25. U.S. Patent Office, patent number 587,649.

26. "Getting to Death: Are Executions Constitutional?," pp. 625–26.

27. "Two Shocks Were Needed."

28. Frederick Drimmer, *Until You Are Dead: The Book of Executions in America* (New York: Citadel Press, 1990), p. 25. Eddie Mays was electrocuted on June 15, 1963.

29. "Electricity as a Life-Taker," *New York Morning Sun,* 4 November 1888, p. 3.

30. Harold P. Brown, "The New Instrument of Execution," *North American Review* 149, no. 396 (1889): 586–93.

31. George E. Fell, "The Influence of Electricity on Protoplasm, with Some Remarks on the Kemmler Execution," *The Physician and Surgeon* 12, no. 10 (October 1890): 439–40.

32. Theodore Bernstein, "Effects of Electricity and Lightning on Man and Animals," *Journal of Forensic Sciences,* 18 (1973): 3–11.

33. Carlos F. MacDonald, M.D., "The Infliction of the Death Penalty by Means of Electricity," *New York Medical Journal,* 55 (May 7, 1892): 505–9. Quote from p. 507.

34. Deborah W. Denno, "Is Electrocution an Unconstitutional Method of Execution? The Engineering of Death over the Century," *William and Mary Law Review* 35, no. 2 (winter 1994): 663.

35. Harold Hillman, "An Unnatural Way to Die," *New Scientist* (27 October 1983): 277–79. For an excellent discussion of the affidavits presented in various court challenges see Denno, "Getting to Death: Are Executions Constitutional?," pp. 319–464.

36. Denno, "Is Electrocution an Unconstitutional Method of Execution?," pp. 641–42.

37. Paul A. David, "The Hero and the Herd in Technology History: Reflections on Thomas Edison and the Battle of the Systems." In *Favorites of Fortune: Technology, Growth, and Economic Development Since the Industrial Revolution,* edited by Patrice Higonnet, David S. Landes, and Henry Rosovsky (Cambridge, MA: Harvard University Press, 1991), pp. 72–119.

38. Because of the scheduled presidential elections of 1892, the exposition was planned for 1893.

39. At the instigation of Elbridge Gerry, an electric chair was on display at the world's fair. "Improvements in Electrocution," *Electrical Engineer* (11 January 1893): 36.

40. Westinghouse's solution was to invent the "stopper" lamp. It was called such because the bulb was made in two parts, with the top fitting into the bottom like a cork into a bottle. Although it was of inferior design, it did the trick until Edison's patent expired.

41. Francis E. Leupp, *George Westinghouse: His Life and Achievements* (Boston: Little, Brown, 1919), p. 164.

42. "To Beat a Trust, That Is What the Westinghouse Electric Company Is After," *New York Evening Advertiser*, 29 March 1892. Edison Archives 146/786.

43. Leupp, *George Westinghouse*, pp. 163–64.

44. For more information on Niagara Falls see *Niagara Falls: Its Power Possibilities and Preservation*, Washington, D.C.: The Smithsonian Institution, 15 January 1925.

45. Henry G. Prout, *A Life of George Westinghouse* (New York: American Society of Mechanical Engineers, 1921), p. 146.

46. For a brief description of events, see Matthew Josephson, *Edison: A Biography* (New York: McGraw-Hill, 1959; reprinted New York: John Wiley & Sons, 1992), pp. 339–63.

47. Thomas Edison to Heny Villard, 24 February 1890 and 1 April 1889; Villard Papers, Houghton Library, Harvard University. Quoted in Josephson, *Edison: A Biography*, p. 361.

48. Quoted in Josephson, *Edison: A Biography*, p. 366.

49. David, pp. 92–93.

50. Leonard S. Reich, *The Making of American Industrial Research* (Cambridge: Cambridge University Press, 1985), p. 52.

51. For a brief discussion of how these standards came about, see Thomas P. Hughes, *Networks of Power: Electrification in Western Society, 1880–1930* (Baltimore: Johns Hopkins University Press, 1983), pp. 127–28.

52. Thomas Edison, "The Storage Battery and the Motor Car," *North American Review* 175, no. 1 (July 1902).

53. Neil Baldwin, *Edison: Inventing the Century* (New York: Hyperion, 1995), pp. 346, 298.

54. For a description of the museum, see Harold K. Skramstad Jr. and Jeannie Head, *An Illustrated History of the Henry Ford Museum and Greenfield Village* (Dearborn, Mich.: 1990).

55. The Edison National Historic Site is located on Main Street and Lakeside Avenue, West Orange, New Jersey.

56. See www.georgewestinghouse.com/george.html.

57. Up-to-date information on General Electric can be found at www.ge.com.

58. Up-to-date information on Westinghouse can be found at www.westinghouse.com.

59. "Brown, Harold Pitney," *National Encyclopaedia of American Biography* (New York: James T. White & Company, 1930), pp. 329–30.

60. Brown's letterhead read "Consulting Engineer, Expert in Patent Applications and Research, Expert in Catalogue or Descriptive Writing, Expert in Developing Inventions, Maker of Electrical Contact Alloys, The Concrete Atomizer." Letterhead also listed him as a member of the American Society of Mechanical Engineers.

61. Harold P. Brown to Thomas A. Edison, 10 February 1927. Edison Archives.

62. Elsewhere Brown further described his invention as "an electrochemical method of removing from the human system poisons due to fatigue, old age and metabolism and of introducing the necessary tonic or nerve food, and electrochemical methods of internal cleansing and of detecting and locating internal congestion." Quoted in James F. Penrose, "Inventing Electrocution," *American Heritage of Invention and Technology* 10 (spring 1994): 44.

63. Harold P. Brown to Thomas A. Edison.

64. William H. Meadowcraft to Harold P. Brown, 19 March 1927. Edison Archives.

65. William H. Meadowcraft to Harold P. Brown, 17 May 1927. Edison Archives.

66. Eddie Mays was electrocuted on June 15, 1963.

67. Drimmer, *Until You Are Dead*, p. 45.

68. "Chair Was Destroyed in 1929 Riot," *Auburn Citizen*, 6 August 1990, p. A8.

69. Denno, "Is Electrocution an Unconstitutional Method of Execution?," p. 619.

70. Ibid., 591–92.

71. *In re Kemmler*, 7 N.Y.S. 145 (Cayuga County Ct., 1889). OR The People of the State of New York, ex rel. William Kemmler, against Charles F. Durston, as warden of the state prison at Auburn, N.Y., p. 447.

72. *Furman v. Georgia* (408 U.S. 238). This decision was actually a collective decision of three cases: *Furman v. Georgia, Jackson v. Georgia,* and *Branch v. Texas.*

73. Oklahoma was the first state to adopt lethal injection in 1977. It was viewed by many as a more humane means of execution than electrocution. The first man executed by lethal injection was Charlie Brooks in Texas on December 2, 1982.

74. For up-to-date information on the methods of execution in use and the number of people on death row in the various states, see www.deathpenaltyinfo.org.

INDEX

Italicized page numbers indicate illustrations.

DEAD MAN WALKING
by Sister Helen Prejean

In 1982, a Roman Catholic nun became the spiritual advisor to a convicted murderer. Powerfully and persuasively, with a compassion that embraces not only the terrified killer but also the families of his victims and the men who executed him, Sister Helen Prejean narrates Patrick Sonnier's walk to the electric chair. Confronting both the needs of a crime-ridden society and the Christian imperative of love, *Dead Man Walking* is a gripping and profound spiritual journey.

Memoir/Current Affairs/0-679-75131-9

THE EXECUTIONER'S SONG
by Norman Mailer

This Pulitzer Prize–winning book follows the short, blighted career of Gary Gilmore, an intractably violent product of America's prisons who became notorious for two reasons: first, for robbing and killing two men in cold blood, and second, for insisting on dying for his crime. To do so, he had to fight a system that seemed paradoxically intent on keeping him alive long after it had sentenced him to death. *The Executioner's Song* is a trip down the wrong side of the tracks to the deepest sources of American loneliness and brutality.

Current Affairs/0-375-70081-1

CROSSED OVER
by Beverly Lowry

When the novelist Beverly Lowry came across a newspaper story about Karla Faye Tucker, the infamous Houston murderer, she was struck by Tucker's innocent beauty, the stunning brutality of her crimes—committed with a pickax—and the stories of her spiritual awakening on death row. Intrigued by these apparent contradictions, Lowry began to visit her in prison and formed a profound and genuine friendship with Tucker, who ten years later would become the first woman to be executed in Texas since 1863.

Memoir/0-375-71380-8